T0140225

Advances in Experimental Medicine and Biology

Volume 1135

More information about this series at http://www.springer.com/series/5584

Avia Rosenhouse-Dantsker • Anna N. Bukiya

Editors

Direct Mechanisms in Cholesterol Modulation of Protein Function

 Springer

Editors
Avia Rosenhouse-Dantsker
Department of Chemistry
University of Illinois
Chicago, IL, USA

Anna N. Bukiya
Department of Pharmacology
The University of Tennessee HSC
Memphis, TN, USA

ISSN 0065-2598   ISSN 2214-8019   (electronic)
Advances in Experimental Medicine and Biology
ISBN 978-3-030-14267-4   ISBN 978-3-030-14265-0   (eBook)
https://doi.org/10.1007/978-3-030-14265-0

This Springer imprint is published by the registered company Springer Nature Switzerland AG
The registered company address is: Gewerbestrasse 11, 6330 Cham, Switzerland

Preface

Cholesterol is a major lipid component of the plasma membrane where it constitutes up to ~50 mol% of the total membrane lipids. As such, it is not surprising that cholesterol effects on protein function have been historically attributed to its effect on the physical properties of lipid bilayers. First indications that cholesterol may bind directly to proteins started to emerge in the mid-1970s in studies on the eukaryotic sugar transport system [1], the Folch-Lees proteolipid protein, the major protein component of myelin [2], the Na^+/K^+-ATPase [3], and the band 3 protein that constitutes the main integral protein of the human erythrocyte membrane [4]. These studies opened a floodgate, and since then, cholesterol has been shown to play a direct role in the regulation of an ever-growing number of proteins.

In 1998, the first amino acid consensus sequence for cholesterol binding, the cholesterol recognition amino acid consensus (CRAC) motif, was proposed in the context of the peripheral-type benzodiazepine receptor, a transmembrane protein that mediates the translocation of cholesterol [5]. While the CRAC motif has been identified in multiple proteins since then, several other cholesterol binding motifs followed, and the characteristics of cholesterol-binding sites in proteins have continued to be defined.

The first structural evidence that cholesterol can bind directly to proteins emerged in mid-2002 with the determination of the structure of the cryptogein-cholesterol complex via X-ray crystallography at a 1.45 Å resolution [6]. Despite its small size, cryptogein, a fungal elicitor, displayed a large inner hydrophobic cavity that harbored the cholesterol molecule. The same year, a structure of the ligand-binding domain of the retinoic acid orphan receptor α was determined at 1.63 Å resolution in complex with cholesterol [7]. This was followed by a structure of the cholesterol-bound oxysterol-binding protein Osh4 at a 1.6 Å resolution in 2005 [8].

In 2007, a structure of the β2 adrenergic G-protein-coupled receptor was crystallized at a 2.4 Å resolution in complex with cholesterol [9]. In this structure, cholesterol mediated receptor-receptor interactions improving the stability of the receptor. This was another milestone in the quest to uncover the direct roles of cholesterol-protein interactions in protein function.

Over the course of several decades, numerous functional, structural, and computational studies have continued to shape our understanding of cholesterol-protein interactions, unraveling the growing number of roles that they play in cellular function. These range from cholesterol transport and storage to protein stability, folding, and localization. While many questions regarding the underlying molecular mechanisms remain unresolved, significant advances in our understanding of direct cholesterol-protein interactions have been made in recent years, and are the topic of this volume.

This is the second in a sequel of two volumes on the mechanisms of cholesterol modulation of protein function. The first volume (1115 in the Advances in Experimental Medicine and Biology Series) focused on sterol specificity as a means to distinguish between direct and indirect effects of cholesterol as well as on indirect mechanisms that impact protein function in response to variations in cholesterol level. The current volume complements the picture by focusing on protein targeting via direct interactions of the cholesterol molecule with sterol-sensing protein sites.

The first part of this volume introduces the reader to the general characteristics of cholesterol binding sites. This part starts with a survey of the different cholesterol-binding motifs that have been proposed over the years followed by an overview of the major classes of proteins that bind steroids and the insights gained from their study using X-ray crystallography. It then continues to two studies that utilize the growing number of structures of cholesterol-bound proteins available in the Protein Data Bank to present new insights into the molecular and structural characteristics of cholesterol-binding sites. The second part of this volume delves into more specific cases of cholesterol binding to G-protein-coupled receptors, ion channels, and cholesterol transporters that have been studied using combinations of experimental and computational approaches.

The editors are grateful to all the authors who contributed to this project aimed at portraying the intricate interactions between a variety of proteins and cholesterol. The editors are also thankful to senior mentors, collaborators, and colleagues for stimulating discussions, and for fostering a supportive environment for the completion of this diverse collection of contributions to the field.

Chicago, IL, USA   Avia Rosenhouse-Dantsker
Memphis, TN, USA   Anna N. Bukiya

References

1. Komor B, Komor E, Tanner W. Transformation of a strictly coupled active transport system into a facilitated diffusion system by nystatin. J Membrane Biol. 1974;17:231–8.
2. London Y, Demel RA, Geurts Van Kessel WSM, Zahler P, Van Deenen LLM. The interaction of the "Folch-Lees" protein with lipids at the air-water interface. Biochim Biophys Acta. 1974;332:69–84.
3. Giraud F, Claret M, Garay R. Interactions of cholesterol with the Na pump in red blood cells. Nature. 1976;264:646–8.
4. Klappauf E, Schubert D. Band 3-protein from human erythrocyte membranes strongly interacts with cholesterol. FEBS Lett. 1977;80:423–5.
5. Li H, Papadopoulos V. Peripheral-type benzodiazepine receptor function in cholesterol transport. Identification of a putative cholesterol recognition/interaction amino acid sequence and consensus pattern. Endocrinology. 1998;139:4991–7.
6. Lascombe M-B, Ponchet M, Venard P, Milat M-L, Blein J-P, Prangé T. The 1.45 Å resolution structure of the cryptogein-cholesterol complex: a close-up view of a sterol carrier protein (SCP) active site. Acta Crystallogr D Biol Crystallogr. 2002;58:1442–7.
7. Kallen JA, Schlaeppi JM, Bitsch F, Geisse S, Geiser M, Delhon I, Fournier B. X-ray structure of hRORalpha LBD at 1.63 Å: structural and functional data that cholesterol or a cholesterol derivative is the natural ligand of RORalpha. Structure. 2002;10(12):1697–702.
8. Im YJ, Raychaudhuri S, Prinz WA, Hurley JH. Structural mechanism for sterol sensing and transport by OSBP-related proteins. Nature. 2005;437:154–8.
9. Cherezov V, Rosenbaum DM, Hanson MA, Rasmussen SG, Thian FS, Kobilka TS, Choi HJ, Kuhn P, Weis WI, Kobilka BK, Stevens RC. High-resolution crystal structure of an engineered human beta2-adrenergic G protein-coupled receptor. Science. 2007;318:1258–65.

Contents

Part I
General Characteristics of Cholesterol Binding Sites

Cholesterol-Recognition Motifs in Membrane Proteins

Jacques Fantini, Richard M. Epand, and Francisco J. Barrantes

Abstract The impact of cholesterol on the structure and function of membrane proteins was recognized several decades ago, but the molecular mechanisms underlying these effects have remained elusive. There appear to be multiple mechanisms by which cholesterol interacts with proteins. A complete understanding of cholesterol-sensing motifs is still undergoing refinement. Initially, cholesterol was thought to exert only non-specific effects on membrane fluidity. It was later shown that this lipid could specifically interact with membrane proteins and affect both their structure and function. In this article, we have summarized and critically analyzed our evolving understanding of the affinity, specificity and stereoselectivity of the interactions of cholesterol with membrane proteins. We review the different computational approaches that are currently used to identify cholesterol binding sites in membrane proteins and the biochemical logic that governs each type of site, including CRAC, CARC, SSD and amphipathic helix motifs. There are physiological implications of these cholesterol-recognition motifs for G-protein coupled receptors (GPCR) and ion channels, in membrane trafficking and membrane fusion (SNARE) proteins. There are also pathological implications of cholesterol binding to proteins involved in neurological disorders (Alzheimer, Parkinson, Creutzfeldt-Jakob) and HIV fusion. In each case, our discussion is focused on the key molecular aspects of the cholesterol and amino acid motifs in membrane-embedded regions of membrane proteins that define the physiologically relevant crosstalk between the two.

J. Fantini (✉)
INSERM UMR_S 1072, Marseille, France

Aix-Marseille Université, Marseille, France

R. M. Epand
Department of Biochemistry and Biomedical Sciences, McMaster University, Health Sciences Centre, Hamilton, ON, Canada

F. J. Barrantes
Laboratory of Molecular Neurobiology, Biomedical Research Institute (BIOMED), UCA–CONICET, Buenos Aires, Argentina

© Springer Nature Switzerland AG 2019
A. Rosenhouse-Dantsker, A. N. Bukiya (eds.), *Direct Mechanisms in Cholesterol Modulation of Protein Function*, Advances in Experimental Medicine and Biology 1135, https://doi.org/10.1007/978-3-030-14265-0_1

Our understanding of the factors that determine if these motifs are functional in cholesterol binding will allow us enhanced predictive capabilities.

Keywords Cholesterol · Binding site · Membrane protein · Membrane fusion · Virus fusion · Neurological disease

1 Overview of Lipid Recognition Motifs in Proteins: Range of Specificity and Affinity

Quantifying binding affinities in interactions between membrane components or between a membrane component and a water-soluble molecule can be far from straightforward, since the membrane components may be in a different, 2-dimensional phase, meaning that their binding cannot be dealt with by applying the same methods as those used in solution thermodynamics. Qualitative binding behavior, however, can be more easily assessed. Binding specificity for membrane lipid components often depends on interaction with the lipid headgroup. For example, phosphatidylinositol and its several phosphorylated derivatives have very different binding affinities for certain proteins determined by the number and position of phosphate groups on the inositol ring. This type of headgroup structure, with its capacity to form hydrogen and electrostatic bonds, does not exist for sterols. Cholesterol, for example, has only a single OH group as its polar moiety. In addition to the headgroup, however, binding can also occur at the hydrocarbon portion of the lipid, accounting for the observation that both the headgroup and hydrocarbon regions of lipids determine their biological function [1].

In addition to the direct binding of proteins to cholesterol, cholesterol can also induce the binding of proteins to membranes by affecting membrane physical properties. Cholesterol plays important roles in the formation of domains in biological membranes [2], as well as in modulating membrane physical properties [3]. Because of the importance of cholesterol in determining membrane properties, there are multiple mechanisms involving cholesterol binding to proteins, to maintain cholesterol homeostasis [4]. This regulation of the metabolism and transport of cholesterol is dependent on the specific cholesterol binding sites on proteins. The specificity of protein binding to cholesterol will likely include interactions with both the hydroxyl group and with portions of the hydrocarbon region. The degree of specificity can be assessed by comparing the binding to cholesterol with binding to ergosterol, a closely related sterol from yeast. Stereochemical isomers of cholesterol can also test specificity [5]. The sterol analogs include epicholesterol, the 3′ epimer of cholesterol and *ent*-cholesterol, the enantiomer of cholesterol. *Ent*-cholesterol is the closest analog, but its use requires the total synthesis of the sterol. With epicholesterol the hydroxyl group protrudes from the sterol ring system at an angle in contrast with cholesterol in which the sterol ring system will be in the same plane as the hydroxyl group. Hence, it is not likely that a protein binding site for cholesterol would also bind epicholesterol. The situation is different with *ent*-cholesterol, the enantiomer or mirror image of cholesterol. Lipids generally have few chiral sites, so that the interactions of cholesterol and *ent*-cholesterol with phospholipids in bilayer membranes are generally identical. However, in the presence of peptides or proteins, there are chiral sites at every amino acid residue, with the result that there is usually a difference between the binding of cholesterol vs. *ent*-cholesterol [6], though there are examples of proteins that can bind equally well to cholesterol and *ent*-cholesterol. Differences in the binding affinity of these two enantiomorphs can therefore be used as evidence of the presence of a cholesterol binding site in proteins, whereas if the binding affinities are the same one may conclude that the cholesterol binding site in the protein is not stereospecific. There has been limited use of this tool since *ent*-cholesterol is not commercially available and its synthesis is complex.

Another factor affecting protein binding to a lipid in a membrane is the distribution in the plane of the membrane and the formation of domains. This is particularly true for cholesterol, which can promote the formation of phases showing liquid-liquid immiscibility. The liquid-ordered, L_o phase has a higher cholesterol concentration [7]. Such cholesterol-enriched phases have been suggested to represent putative "raft" phases that occur naturally in biological membranes. Thus, another factor potentially affecting protein binding to cholesterol in membranes containing liquid-ordered domains, is whether or not the protein sequesters into these domains. Because the mol fraction of cholesterol is higher in these domains, proteins will not require such a high affinity to bind cholesterol.

In many cases, the interaction of cholesterol with proteins may be even more complicated than a single uniform binding site, as described above. For example, an NMR study of the interaction of cholesterol with the β_2 adrenergic receptor showed that there were two classes of cholesterol binding to this protein. One class corresponded to a limited number of high affinity sites having sub-nanomolar affinity for this lipid. However, there was a second class of cholesterol binding in fast exchange with unbound cholesterol and with an affinity that was lower by several orders of magnitude. It was suggested that these represented transient cholesterol clusters around high affinity cholesterol binding sites [8]. There has also been a recent molecular dynamics study demonstrating distinct cholesterol binding sites in the A_{2A} adenosine receptor [9].

2 Cholesterol-Recognition Motifs

Studying a lipid-protein binding process calls for an understanding of the basic principles of this interaction. In the case of cholesterol and membrane proteins, the problem may look simple at first glance, but as we will see, it can be far more complex than expected. Schematically, the binding reaction involves two partners: a cholesterol molecule and a membrane protein. Since the lipid bilayer of the biological membrane is the natural medium for the cholesterol molecule, several simplifications can reasonably be applied to the system. Firstly, only protein domains that cross the lipid bilayer are involved. Although this may be considered patently obvious, exceptions to this rule have been reported, as for the human oncoprotein Smoothened (SMO), which displays a functional cholesterol binding site in the extracellular domain, i.e. outside the membrane bilayer boundaries [10]. In the case of human phospholipid scramblase 1, cholesterol binds to a specific domain that includes both a membrane-embedded and an extracellular coil [11]. Apart from these rare cases, most cholesterol binding sites of integral membrane proteins lie within their α-helical transmembrane domains (TMDs) that totally cross the lipid bilayer. Several cholesterol-binding sites have been found in TMDs [12, 13]. Some of these sites are clearly three-dimensional [14, 15], whereas others follow linear motifs [16, 17]. Among these motifs, the linear CRAC domain (Cholesterol Recognition/interaction Amino acid Consensus sequence) [18] and its reverse formulation CARC [19] have received considerable attention.

2.1 CRAC Motif

The CRAC motif is defined by the consensus (L/V)-X_{1-5}-(Y)-X_{1-5}-(K/R) from the N-terminus to C-terminus direction [18]. This motif can be considered a chemical fingerprint of cholesterol. Each of the three amino acid residues that define the CRAC motif has a specific function in cholesterol recognition. The N-terminal branched residue (valine or leucine) binds the iso-octyl chain of cholesterol through van der Waals interactions. At the opposite end, the C-terminal polar residue (lysine or arginine) faces the OH group of cholesterol, allowing the establishment of a hydrogen bond. In addition, the CRAC motif is vectorial, imposing a parallel "head-to-head/tail-to-tail" geometry to the CRAC/cholesterol complex (Fig. 1). This, in turn, facilitates the aromatic structure of tyrosine stacking onto one of the four rings of sterane. It should be noted that the position of tyrosine is determined by the length of a couple of X_{1-5} linkers that separate the aromatic residue of CRAC from the ends of the motif. The presence of such variable segments, which differ in both length and composition, has been viewed as a serious weakness by some authors [20]. But in fact, this

Fig. 1 Geometry of the CRAC/cholesterol complex. The motif is oriented in the N-ter (top) to C-ter (bottom) direction. It displays three distinct zones (apolar in blue, aromatic in yellow, cationic in purple) that fit with the chemical structure of cholesterol

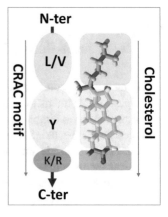

variability reveals a hallmark of cholesterol binding sites found in most cholesterol-TMD complexes: the essential contribution of CH-Pi stacking interactions [21].

When an aromatic ring faces an aliphatic cycle, it adjusts its orientation so that the Pi electron cloud attracts the hydrogen atoms linked to the aliphatic cycle, resulting in a coordinated network of favorable interactions. This particular case of attraction between the C-H groups of a saturated cyclic hydrocarbon and an aromatic ring is referred to as the "stacking CH-Pi interaction" [21]. Sometimes, the induced fit mechanism that directs the respective orientation of both rings results in a near perfect geometry, as shown in Fig. 2. In the case of the CRAC-cholesterol complex, the establishment of such an optimal geometry requires that the aromatic ring of Tyr is parallel to sterane. Obviously, it is the distance between Tyr and the ends of the motif that determines which of the four rings of cholesterol is selected for the establishment of the stacking CH-Pi system. Thus, the length of the linkers (from one to five amino acid residues) allows several possible stacking interactions. In other words, thanks to both linkers, the Tyr residue can be viewed as a cursor able to occupy any possible position in the motif [12], and this unique feature would not be possible if the linkers had a fixed length. The total length of the CRAC motif ranges from five amino acid residues (both linkers with only one residue) to 13 residues (both linkers with five residues). The maximal size of CRAC motifs is by no means a coincidence. Indeed, an α-helix stretch of 13 amino acid residues has approximately the same size as cholesterol, i.e. 20 Å [22, 23]. The fact that the linkers have no sequence requirements confirms that only their length matters, which is remarkably consistent with the biochemical mechanisms underlying the formation of a CRAC-cholesterol complex.

In membrane areas where cholesterol is present in both leaflets of the plasma membrane, the same TM domain can theoretically interact with two cholesterol molecules (one in each leaflet). However, the vectorial nature of the CRAC motif is compatible with only one of these possibilities, depending on the orientation of the TMD. If the TMD crosses the bilayer in the N-terminus to C-terminus direction, the CRAC domain may interact with a cholesterol molecule located in the cytoplasmic

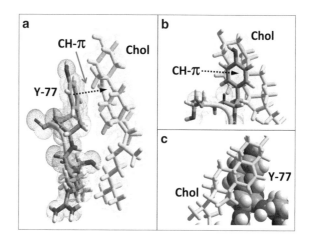

Fig. 2 CH-π stacking interaction in the CRAC/cholesterol complex. Three distinct views of cholesterol (in yellow) (**a–c**) bound to the CRAC domain of the human delta-type opioid receptor are shown. The near perfect superposition of the aromatic ring of Tyr-77 onto the second ring of cholesterol is particularly well illustrated in **b** and **c**

leaflet of the membrane, but not in the extracellular leaflet [23]. Therefore, a CRAC domain in the unique TMD of a bitopic membrane protein will interact with cholesterol in the inner leaflet. Similarly, TMDs I, III, V, and VII of G-protein coupled receptors (GPCRs) displaying a CRAC motif will also select cholesterol in the inner leaflet [24]. Conversely, the interaction of CRAC with cholesterol in the exofacial leaflet requires that the TMD crosses the bilayer in the C- terminus to N-terminus direction. This kind of situation applies for type II bitopic membrane proteins and GPCRs (TMDs II, IV, and VI).

The CRAC motif has been found in various proteins known to bind cholesterol and in many cases the interaction between cholesterol and CRAC has been confirmed by various physicochemical and/or functional approaches [12, 24–28]. Moreover, single mutations in the CRAC domain have been found to markedly decrease or even abolish the interaction. In this respect, it should be noted that in most instances, the Tyr residue cannot be replaced by Phe or Trp [29–31]. Nevertheless, a thorough analysis of CRAC domains through molecular docking studies suggests that, at least in some cases, the aromatic residue may not be directly involved in cholesterol recognition [13]. In other cases, the aromatic ring of Phe could sustain CH-Pi stacking interactions when Tyr is not present in the motif [16]. Future studies will likely lead to a refinement of the definition of the CRAC domain, especially for membrane proteins.

2.2 CARC Motif

The impossibility of the CRAC motif to interact with cholesterol in the exofacial domain of a large number of TMDs implied the possible existence of another specific cholesterol-binding motif. Indeed, the discovery of a new motif, referred to as CARC, was primarily due to the fact that no CRAC motifs were found in the TMDs of the nicotinic acetylcholine receptor protein; instead, CARC motifs were found [19]. Basically, CARC is an inverted and slightly modified version of the CRAC motif: $(K/R)-X_{1-5}-(Y/F/W)-X_{1-5}-(L/V)$. The CARC domain displays remarkably specific features that take into account the membrane environment. Firstly, the central residue is still aromatic, but unlike CRAC which, in theory, has a specific requirement for Tyr, the CARC motif can accept Tyr, Phe, or Trp, consistent with the presence of all these residues in TMDs of various membrane proteins [32]. Secondly, the basic amino acid of CARC is located at the N-terminus. This distinctive feature explains why the CARC domain of class I membrane proteins (the most abundant bitopic proteins) can form a complex with cholesterol in the exofacial leaflet (Fig. 3). The same is true for TMDs I, III, V, and VII of GPCRs.

The biochemical rules that apply to the CRAC-cholesterol interactions also apply for CARC, since both motifs share a similar organization, i.e. a triad of mandatory amino acids with a central aromatic residue flanked by a basic and a branched apolar residue at each end. In both cases, spacers consisting of one to five unspecified amino acids ensure that the aromatic ring in the central position of the cholesterol-binding motif can optimally stack onto one of the sterane rings.

The CARC domain has been detected in a wide range of membrane proteins, including neurotransmitter receptors and transporters, ion channels and GPCRs [12, 13, 15, 16, 24, 33–35]. The nicotinic acetylcholine receptor displays 15 cholesterol binding sites (3 per subunit) that fulfill the CARC algorithm [19]. Docking studies have led to the proposal of a crown-like distribution of those cholesterol molecules around the receptor (Fig. 4), in agreement with the early views stemming from elec-

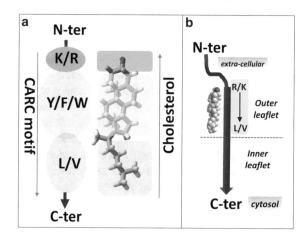

Fig. 3 The CARC motif of Class I bitopic membrane proteins is located in the outer leaflet of the plasma membrane. The CARC motif and cholesterol are represented with the same color as in Fig. 1. (**a**) Topology of the CARC-cholesterol complex. (**b**) Membrane localization of the CARC-cholesterol complex. The border between the outer and inner membrane leaflets is indicated by a dashed line

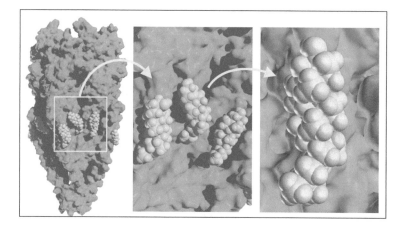

Fig. 4 Docking of 15 cholesterol molecules onto the nicotinic acetylcholine receptor. Three cholesterol molecules bound to the γ subunit of the acetylcholine receptor are shown at a different scale. The picture on the right shows the cholesterol molecule bound to the CARC motif in the fourth TMD of the γ subunit

tron spin resonance studies [36], as reviewed in ref. [37]. Biophysical studies with synthetic peptides encompassing a CARC domain have provided experimental support to the cholesterol-binding activity of the motif. A deuterium NMR spectrum of the CARC motif of the *Torpedo* nicotinic acetylcholine receptor γ-TM4 showed that the presence of cholesterol within the bilayer induced a reduction in the rotational motion of the peptide within the bilayer, a change consistent with cholesterol promoting the oligomerization of the γ–TM4 segment [16]. Moreover, mutational studies of this domain confirmed the prominent role of its central Phe residue. Indeed, the interaction with a cholesterol-containing monolayer was dramatically decreased by a single Phe→Ala mutation, whereas it was not significantly affected by the conservative Phe→Trp substitution [16]. Consistent with these experimental data, molecular docking studies indicated that the central aromatic residue of this CARC domain (Phe-452) is the most important energetic contributor of the complex.

A TMD has generally 22–26 amino acid residues [16]. Since CARC and CRAC motifs comprise between 5 and 13 amino acid residues, it is theoretically possible for a TMD to possess both motifs. An analysis of sequence databases has recently confirmed that such a "mirror" topology actually exists in various types of membrane proteins, including ion channels, neurotransmitter receptors, ABC transporters and GPCRs [16]. In all these cases, molecular dynamics simulations indicated that mirror TMDs could perfectly well accommodate two cholesterol molecules in a typical tail-to-tail orientation, one bound to CARC and the other to CRAC (Fig. 5). Future studies will be necessary to evaluate the functional impact of two symmetric cholesterol molecules on membrane proteins.

A common criticism of the definition of CRAC and CARC is that the consensus sequence defining the two motifs is too general to have any predictive value with respect to cholesterol binding [20]. Indeed, available crystal structures of membrane proteins complexed with cholesterol have made it possible to identify 3D pockets rather than linear binding sites [38]. Interestingly, the biochemical rules controlling cholesterol binding to these 3D sites are basically the same as those that apply for

Fig. 5 Mirror topology of CARC/CRAC motifs within the same TMD. Three distinct views of the complex are shown. Cholesterol in yellow is bound to CARC, and cholesterol in red is bound to CRAC. The TMD shown is the seventh TMD of the human adenosine receptor A1

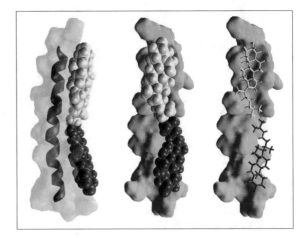

cholesterol binding to CARC or CARC motifs. In particular, the involvement of an aromatic ring that stacks onto the sterane backbone of the sterol seems to be a hallmark of cholesterol-protein interactions in the membrane environment [13]. The particular topology of TMDs together with a universal mechanism of membrane cholesterol binding will probably render possible the prediction of potential cholesterol-binding motifs from sequence databases [16].

2.3 Sterol-Sensing Domains (SSD)

Unlike CRAC and CARC motifs, that comprise protein segments containing 13 amino acid residues or less, the sterol-sensing domain (SSD) is much larger. It contains approximately 180 residues organized as five consecutive transmembrane helices joined by short extramembranous loop regions. Interest in SSDs comes from the fact that they are found in several proteins involved in cholesterol transport, metabolism and storage [39, 40]. An SSD was originally identified in the enzyme 3-hydroxy-3-methylglutaryl coenzyme A-reductase (HMG-CoAR), the enzyme that catalyzes the rate determining step in cholesterol biosynthesis [41]. SSDs have also been found in SCAP (the sterol regulatory element-binding protein-cleavage activating protein). SCAP is an integral membrane protein found in the endoplasmic reticulum that plays a major role in regulating the transcription of genes involved in cholesterol biosynthesis [39, 41]. Other proteins in which SSDs have been identified and which have some relationship to cholesterol include 7-dehydrocholesterol reductase, an enzyme involved in cholesterol biosynthesis, and the Niemann-Pick C1 protein (NCP1), involved in intracellular lipid transport and lipid storage. The most prominent effect of defective NCP1 is the accumulation of unesterified cholesterol in an endosomal/lysosomal compartment. This accumulation occurs because the proteins NCP1 and NCP2 are required for transporting cholesterol out of lysosomes. NCP1 has also been shown to be required for the entry of Ebola virus to the cytoplasm [42, 43]. Other proteins with SSD include Patched (Ptc), a tumor suppressor involved in the signal transduction pathway for Hedgehog, a lipidated protein with covalently-linked cholesterol; Dispatched (DISP), a protein involved in the release of Hedgehog; and PTR, a protein related to Ptc whose function has still not been fully elucidated.

Because of the central role it plays in cholesterol homeostasis, SCAP is the most-studied SSD-containing protein from a mechanistic point of view. It has been shown that the activity of SCAP regulates cholesterol biosynthesis over a low and narrow range of cholesterol concentrations in the endoplasmic reticulum [44], a phenomenon that can be explained by the rapid rise in cholesterol activity over the narrow range of concentrations in which SCAP is activated [44]. Any explanation other than the direct modulation of SCAP activity through the binding of cholesterol to this protein would be hard to justify. However, little is known about the nature of this binding and which region of SCAP is involved in the interaction with cholesterol. There is one study determining the binding of cholesterol to isolated segments of SCAP, suggesting that the cholesterol binding site was in loop 1 [45]. This paper compared the binding of cholesterol and competition with other sterol analogs to loop 1 versus to the entire SSD domain. The study concluded that loop 1 contained the cholesterol binding site, though there are some caveats to this conclusion. In the first place it is not clear how well the protein fragment mimics the structure of this region in the intact protein. Secondly, the binding studies required the addition of a low concentration of detergent and furthermore the binding to loop 1 was done in an all-or-none manner. Given the highly sigmoidal dependence of cholesterol binding, it would be interesting to see a dose-response curve of cholesterol binding to loop 1. It is particularly difficult to determine the specific binding of a lipid, such as cholesterol, to a protein in an insoluble membrane fraction. Other evidence for the involvement of a particular region of SCAP in binding comes from functional studies in which mutations were introduced in the extra-membranous loop 6 of SCAP containing the sequence MELADL. MELADL is required for the binding of SCAP to Sec23/Sec24. Sec23/Sec24 are proteins on the surface of CopII vesicles that escort SCAP from the endoplasmic reticulum to the Golgi, as the initial step in the pathway for the transcriptional regulation of cholesterol biosynthesis. The nature of the conformational change that results in the loss of exposure of MEDADL when cholesterol binds to SCAP has been recently evaluated by means of the susceptibility of SCAP to proteolytic cleavage [46]. These studies connect cholesterol binding to the functioning of SCAP through a conformational change in the latter that determines the exposure of the MEDADL segment. Cholesterol does appear to bind to loop 1 of SCAP, though additional studies are required to shed more light on this feature. In this context it is interesting to note that one of the juxta-membrane segments of loop 1 is the segment from residue 38 to 46 in human SCAP having the sequence LACCYPLLK. This sequence corresponds to a CRAC motif. However, the cited binding studies were done using the fragment of SCAP comprising residues 46–269 and hence not containing the putative CRAC sequence [45]. The contribution of this CRAC segment to the function of the SSD in SCAP remains to be determined.

There have also been photoaffinity labeling studies using photo-reactive derivatives of cholesterol, demonstrating the importance of the amino-terminal region of SCAP for both cholesterol binding and the functioning of SCAP. The cholesterol affinity probe reacts with a region of SCAP that includes the first transmembrane segment of SCAP [47]. A photoaffinity derivative of 25-hydroxycholesterol does not react with SCAP, showing some specificity for the process. It was demonstrated that the same photolabeling with cholesterol could be performed in whole cells and that reaction with the cholesterol affinity probe blocked the processing of SREBP [47].

Although the functional properties of the SSD-containing protein, SCAP, have been extensively investigated, the structure of an SSD domain is best known for other SSD-containing proteins. NPC1 was first purified and shown to bind to cholesterol and other sterols by Goldstein and his group [48]. The specificity of the cholesterol binding site and the region of the protein to which cholesterol binds was studied [49], leading to the conclusion that the loop 1 region is part of the binding site. Curiously, a Q79A mutation that abolishes the binding of [^{3}H]-cholesterol and of [³H]-25-hydroxycholesterol to full-length NPC1, was nevertheless able to restore cholesterol transport to NPC1-deficient Chinese hamster ovary cells. Thus, the sterol binding site on luminal loop-1 is not essential for NPC1 function in fibroblasts. It was suggested that this site might be required for cholesterol transport in other cells where NPC1 deficiency produces more complicated lipid abnormalities [49]. Recent X-ray crystallographic studies with NPC1 have yielded a structure at 3.3 Å resolution [50]. In order to obtain a higher resolution structure, the full length protein had to be cleaved with a protease that removed a fragment of 313 residues from the amino terminus [51]. This fragment was attached to the remainder of the protein by a polyproline flexible arm [52] that precluded crystallization. The position of the missing N-terminal segment in this crystal structure was shown by cryo-electron microscopy to lie on top of the remainder of the protein in only 45% of the particles, suggesting its flexible linkage to the remainder of the protein [53]. In any case it should be noted that the SSD is not a surface binding site for cholesterol but rather an internal cavity that completely wraps cholesterol (Fig. 6).

Some of the proteins with SSD also contain the short sequence YIYF. It has been shown that SCAP [47] and HMG-CoAR [54] require this tetrapeptide fragment to bind to Insig, an anchoring protein of the endoplasmic reticulum. Interestingly, this sequence, YIYF, is also present in other proteins having some interaction with cholesterol [30]; some of these proteins contain CRAC and/or SSD domains but others do not. The direct role of YIYF in cholesterol binding remains to be fully established.

2.4 Amphipathic Helix

There is a recent example of a short peptide segment that is part of an amphipathic helix that controls the cholesterol-mediated turnover of squalene monooxygenase, a rate-limiting enzyme in cholesterol synthesis [55]. Evidence was presented that a 12-residue segment forming part of an amphipathic helix of squalene monooxygenase

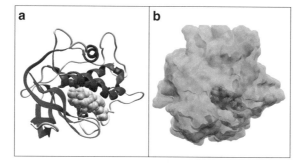

Fig. 6 The SSD of NPC1 totally wraps cholesterol. The protein is represented as a ribbon diagram (**a**) or with a surface rendition (**b**). Alpha helices are in red, beta strands in blue and cholesterol in yellow. The structure of the cholesterol-NPC1 complex is retrieved from PDB file # 3GKI

conveyed cholesterol sensitivity to the binding of the protein to membranes. Although the specific amino acid sequence of this 12-residue fragment may not be required for this cholesterol-dependent function, the general model may have wider applicability. It is known that amphipathic helices have affinity for membranes, but as monomers, they do not insert very deeply into membranes. It is also known that cholesterol promotes tighter packing of membranes. Hence, it is reasonable to suggest that an amphipathic helix that is weakly bound to a membrane may dissociate from the membrane at higher cholesterol concentrations, as was shown in this case [55]. We thus anticipate that other cholesterol-mediated functions will be discovered in the future that depend on the dissociation of amphipathic helices from membranes in the presence of increased cholesterol concentrations.

3 Role of Cholesterol-Recognition Motifs in Binding to GPCR

The possibility that cholesterol could modulate the function of GPCRs has been investigated by numerous authors. In all cases, early studies were faced with the dilemma of being able to decipher what was due to general biophysical effects on the membrane as opposed to specific biochemical effects on receptors. The first account of a direct interaction between cholesterol and a GPCR came from a study on rhodopsin [56]. In these experiments cholestatrienol (a fluorescent sterol) was used to probe interactions between cholesterol and rhodopsin in disk membranes. These interactions were detected by fluorescence energy transfer from protein tryptophan residues to cholestatrienol. The specificity of this interaction was explored by the addition of cholesterol, which inhibited the quenching of fluorescence emission from tryptophan residues of the protein, or ergosterol, which did not. Taken together, these data suggested the existence of a specific cholesterol binding site on rhodopsin [56]. In parallel, other studies were focused on the effect of cholesterol on GPCR function. A pioneering study described the modulatory effect of cholesterol on two GPCRs, the oxytocin receptor and the brain cholecystokinin receptor [57]. Once again, the specificity of cholesterol effects was assessed by comparing its activity with sterol analogues. A major outcome of this study was the demonstration that cholesterol could affect ligand binding to these receptors and subsequent signal transduction [57].

Another way to assess the specificity of cholesterol effects on GPCRs is the use of cholesterol oxidase on native membranes [58]. This enzyme catalyzes the conversion of membrane cholesterol to cholestenone. It turned out that this treatment inhibited the specific binding of agonist and antagonist ligands to the serotonin 5-HT(1A) receptor. Since membrane order was not affected by the enzymatic oxidation of cholesterol to cholestenone, these data suggested that cholesterol could modulate ligand binding to this GPCR through a specific interaction. The definitive evidence for the existence of cholesterol binding sites on GPCRs came from structural data. For a long time, structural studies of GPCRs have been hampered by the lack of reliable crystallization procedures for integral membrane proteins. The advent of the *in meso* technology (also referred to as the lipid cubic phase) has filled this gap, allowing the production of hundreds of X-ray structures of membrane proteins, with GPCRs representing the highest proportion [59]. Interestingly, the addition of cholesterol in a monoacylglycerol matrix has proved to be critical to the production of structure-grade crystals of most membrane proteins, especially GPCRs [60]. As a consequence, GPCRs are often co-crystallized with cholesterol. Although these data confirmed that GPCRs can bind cholesterol, it has not been possible to determine a unique, consensus profile for the cholesterol binding sites observed in these structures. A canonical motif referred to as CCM was detected as a specific cholesterol binding site in the β_2 adrenergic receptor, but not in other GPCRs sharing the same motif [15, 61]. Moreover, two vicinal cholesterol molecules are bound to this receptor, as shown in Fig. 7. Nevertheless, several common features emerged from these structural studies. Consistent with the rules derived from the CRAC/CARC algorithms, branched amino acid residues (Val, Leu, but also Ile) were often involved in cholesterol binding. Stacking interactions mediated by an aromatic residue, including Trp [62], were also frequent. The polar OH group of cholesterol was localized near the water-membrane interface with potential hydrogen bonding to Lys, Arg, but also Asp residues. In fact, it is quite easy to explain the molecular mechanisms of cholesterol-GPCR interactions in the crystal structures obtained by the *in meso* method, but more of these X-ray structures are required before a reliable prediction method for cholesterol-binding sites can be proposed. Meanwhile, identification of CRAC/CARC motifs still represents a valuable strategy to categorize potential points of contact between GPCR TMDs and cholesterol [16].

From a functional point of view, it has been proposed that cholesterol-receptor interactions can exert two complementary effects: (1) increasing the compactness of

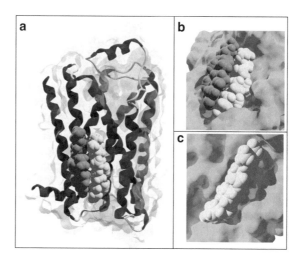

Fig. 7 Two cholesterol molecules bound to the human β_2 adrenergic receptor (retrieved from PDB file # 3D4S). Three distinct views of the complex (**a–c**) are shown. One cholesterol is in green, the other one in yellow

the receptor structure, and (2) improving the conformational stability towards active/inactive receptor states [63]. These specific modulations of receptor structure and functions are mediated by non-annular sites which, in contrast with annular sites, bind cholesterol both specifically and with high affinity [64]. Experimental data in favor of the co-existence of different types of cholesterol interactions with GPCRs has been recently obtained by means of a nuclear magnetic resonance study of the β_2 adrenergic receptor [8]. The authors of this study suggested that a cluster of cholesterol molecules could self-organize around the receptor, certain molecules (non-annular) being in slow exchange and others (annular) in fast exchange, with the former contributing to the specific binding of the latter. In this case, both cholesterol pools could co-operate to facilitate the recruitment of the β_2 adrenergic receptor into cholesterol-rich domains and control its oligomerization state [8].

4   Role of Cholesterol-Recognition Motifs in Ion Channels

Oligomerization is also central to ion channel activity since these membrane proteins consist of individual subunits that are nonfunctional by themselves [65]. Transient receptor potential (TRP) channels, including vanilloid (TRVP), canonical (TRPC), and melastin (TRPM) TRP channels are localized in lipid rafts and are highly sensitive to cholesterol, which controls both their assembly and activity [66]. A thorough study of the effects of cholesterol on ion channel activity has been performed on the inwardly rectifying K^+ channels (Kir) [14, 67, 68]. In these experiments, two stereochemical variants of cholesterol, i.e. *ent*-cholesterol (the cholesterol enantiomer), and epi-cholesterol (which has the distinct orientation of the OH group) were tested and compared with natural cholesterol in functional studies of ion channel activity. Surprisingly, both cholesterol and its chiral isomer were found to bind to the same site through a non-stereospecific mechanism [68]. However, only natural cholesterol could modulate ion channel activity, indicating that sterol binding alone is not sufficient to regulate the channel. From a molecular point of view, the structural determinant of the cholesterol-binding domains displayed by Kir channels, i.e. a hydrophobic pocket [14], is consistent with non-stereoselective binding of sterols through poorly discriminant van der Waals interactions. In another study performed on the nicotinic acetylcholine receptor, epicholesterol was able to substitute for cholesterol in terms of its functional effect [69].

   On the basis of all these data, it is still difficult to specify exactly how cholesterol binding to ion channels controls subunit assembly and/or channel opening probability [70]. Recent studies of amyloid pore formation in the plasma membrane of brain cells have given some clues on the molecular mechanisms controlling the assembly of oligomeric Ca^{2+} channels [71, 72]. Amyloid proteins are generally assumed to self-aggregate into fibers that form large plaques in the brain of patients with neurodegenerative disorders such as Alzheimer, Parkinson, or Creutzfeldt-Jakob diseases [73, 74]. However, healthy individuals may also display significant amounts of amyloid plaques in their brain, so that there is no clear-cut correlation between these deposits and neurological symptoms [75]. In fact, amyloid proteins also form a variety of small neurotoxic oligomers, including amyloid pores which are a particular class of Zn^{2+}-sensitive Ca^{2+} channels [76]. These oligomers are considered to be the most toxic species of amyloid proteins and there is growing evidence that they are closely associated with the pathogenesis of neurodegenerative diseases [77]. The oligomerization process that leads to the formation of amyloid pores is a universal two-step mechanism involving successively a ganglioside and cholesterol [78]. The ganglioside ensures the initial adhesion of the amyloid protein to a lipid raft domain [79]. The insertion of the protein within the plasma membrane is then dependent upon cholesterol which interacts with a specific cholesterol-binding domain displayed by the amyloid protein [80]. The cholesterol binding site of amyloid proteins is linear but is not necessarily a CRAC or a CARC domain. The most important feature of this particular class of cholesterol binding domains is that once inserted in the plasma membrane, they adopt a tilted orientation with respect to the main axis of cholesterol [80], just as viral fusion peptides do [81]. In the case of Alzheimer's β-amyloid peptide, this particular geometry facilitates the oligomerization process which depends on the strict alignment of Lys and Asn residues belonging to vicinal peptide monomers [71, 72]. The assembly of the oligomeric pore is driven by the formation of a hydrogen bond between those Lys and Asn residues [71, 72]. The implication of cholesterol in this process is confirmed by the lack of formation of amyloid pores in cholesterol-depleted cells [78]. Whether cholesterol could play a similar role in larger ion channels remains to be established.

5   Role of Cholesterol-Recognition Motifs in Cholesterol Trafficking

It was shown earlier on that the enzyme that catalyzed the rate determining step in cholesterol biosynthesis, 3-hydroxy-3-methylglutaryl coenzyme A-reductase (HMG-CoAR), contained an SSD. This enzyme is the target for statin drugs. In addition, SCAP (the sterol regulatory element-binding protein-cleavage activating protein) is a protein of the endoplasmic reticulum that controls the transcription of genes involved in cholesterol biosynthesis, through a feedback mechanism involving the binding of cholesterol to an SSD of SCAP. In addition to controlling the biosynthesis of cholesterol, the intracellular cholesterol trafficking proteins, NCP1 and NCP2, also have SSD domains. In addition to these mechanisms, the transport of cholesterol across the plasma membrane of a cell is another function modulated by the level of cholesterol through the binding to specific sites on these transport proteins.

ATP binding cassette (ABC) transporters are a large family of integral membrane protein transporters with homologous structures comprising 6 or 12 transmembrane helices and one or two ATP binding sites. These transporters are subdivided into seven subfamilies based on their structural similarities [82]. One of these subfamilies is the ABCG group. Particular attention has been given to ABCG1, which appears to play a prominent role in the export of cholesterol from cells to HDL and is therefore important in "reverse cholesterol transport", i.e. the movement of cholesterol from peripheral tissues to the liver. Other members of the ABCG subfamily can also transport cholesterol but may operate by an alternative mechanism, since the other lipids they transport are different from ABCG1 [26]. It has been demonstrated that the final transmembrane segment is important for cholesterol transport [26]. This segment contains several CRAC and CARC segments. In particular, mutational analysis has shown that mutation of the CRAC segment containing Y667 results in loss of cholesterol transport to HDL and loss of stability of the protein in the presence of cholesterol [26]. Another ABC transporter that transports cholesterol is ABCA1. However, this transporter does not have any CRAC segments. It is possible that the specificity of transport of ABCA1 comes about because of its specific binding to HDL [83], which is not required for ABCG1 [26]. It should also be kept in mind that the conformation and activity of ABC cassette proteins are influenced by the surrounding lipid [84], so that cholesterol may modulate the activity of these proteins without directly binding to them.

There has recently been a report suggesting a cholesterol transport role for a family of mammalian proteins that are homologous to the ChUP proteins of *C. elegans* [85]. Evidence is presented that these mammalian SIDT proteins transport cholesterol. Furthermore, they have a CRAC domain, which when mutated prevents FRET between these proteins and the fluorescent cholesterol analog, dehydroergosterol [85]. Further studies are required to verify whether these proteins are cholesterol transporters in mammals.

6   Role of Cholesterol in Membrane Fusion

Membrane fusion is an important function in many biological systems. Processes such as the exocytosis of endocytic vesicles, sperm-egg fertilization, cell-cell fusion in bone and heart, infection by enveloped viruses and others, all involve the merging of one membrane with another, promoted by specific proteins, among which lipid plays an important role [86]. There are likely to be some common elements among the various types of membrane fusion in terms of how they are modulated by the lipid environment, including the presence of cholesterol.

There are several mechanisms by which cholesterol may affect the rate of fusion. Cholesterol may be required to bind to a fusion protein to stimulate its fusion activity, it may recruit protein components to the site of fusion so that these proteins are more concentrated in a specific domain of a membrane, cholesterol may modify the biophysical properties of the membrane to favor membrane fusion and/or to stabilize regions of high curvature in fusion intermediates. These putative roles of cholesterol are not mutually exclusive, and a specific membrane fusion process may involve more than one of these properties.

SNARE proteins are required for exocytosis, facilitating the fusion between endocytic vesicles and the plasma membrane. Many of the SNARE proteins required for exocytosis contain CRAC or CARC segments [87]. In addition, it is known that cholesterol is required for exocytosis in neurons [88, 89], endocrine [90], neuroendocrine cells [91, 92] as well as cortical vesicles from sea urchins [93, 94]. However, the role of cholesterol in SNARE-dependent exocytosis does not appear to involve the binding of cholesterol to the SNARE protein, but rather is dependent on the changes cholesterol makes in the physical properties of the membrane and its domain organization [3]. Nevertheless, CARC and CRAC motifs are found in some, although not all, SNARE proteins [87], though there is no evidence that when these domains are present they bind cholesterol or facilitate membrane fusion.

Cholesterol also plays an important role in the fusion of enveloped viruses to cell membranes, with cholesterol-rich domains often serving as the site for such fusion [95], possibly at the interface between the cholesterol-rich domain and the remainder of the membrane [96], as well as for viral assembly and budding [97, 98].

In addition, some viral fusion proteins contain segments that may interact directly with cholesterol. This includes the membrane proximal region of the GP2 protein of Ebola virus that contains the sequence GXXGXXXA, suggested to interact with cholesterol [99]. The sequence GXXXG is often associated with protein dimerization, but this and similar sequences have been shown to also interact with cholesterol in the amyloid precursor protein [100]. This aspect has not been included in the present review among the sequences associated with binding cholesterol owing to the lack of sufficient examples.

One of the most studied CRAC domains associated with viral fusion is the LWYIK segment found in the membrane proximal domain of the HIV fusion protein gp41 [101]. We have shown that the N-acetyl-LWYIK-amide is able to recruit cholesterol into domains in model membranes, resulting in the deeper penetration of the peptide into the membrane [102]. However, in the gp41 fusion proteins of HIV-2 and SIV, in the location of the LWYIK sequence, one finds the modified sequence LASWIK. This is not a CRAC domain, yet these viruses are still active and can undergo membrane fusion [103]. The peptide N-acetyl-LASWIK-amide has less potency than N-acetyl-LWYIK-amide in forming areas enriched in cholesterol. We suggest that the difference between HIV-1 and HIV-2 glycosphingolipid requirements for determining their tropism is related to the difference in their partitioning to cholesterol-rich domains in biological membranes [103]. We tested the stereochemistry of the induction of cholesterol-rich domains by LWYIK and found that substituting cholesterol with its enantiomer, ent-cholesterol, prevented the LWYIK peptide from sequestering cholesterol. However, the enantiomer of N-acetyl-LWYIK-amide, i.e. the peptide with all D-amino acids, was able to segregate cholesterol, indicating that peptide chirality is not required for interaction with cholesterol-containing membranes. However, a specific chirality of membrane lipids is required for peptide-induced formation of cholesterol-rich domains [6]. Computer modeling studies suggested the nature of the non-covalent interactions between cholesterol and the LWYIK peptide. The modeling studies and fluorescence experiments were supported by single residue mutations in the gp41 protein of HIV-1, in which L679 is replaced with I. Despite the similarity of the properties of L and I, this single substitution resulted in a marked attenuation of the ability of JC53-BL HeLa-based HIV-1 indicator cells to form syncytia [31], again suggesting a requirement for a CRAC motif. Mutational studies combined with *in silico* predictions and model system studies of cholesterol clustering, supported a specific model for the interaction of

LWYIK with cholesterol [104, 105]. X-ray scattering studies were carried out to compare the effects of LWYIK and IWYIK on bilayer thickness. With 50% cholesterol, IWYIK was found to decrease the bilayer repeat distance, while LWYIK increased it [106]. There is evidence that longer peptides containing LWYIK may act as inhibitors of HIV fusion activity [107]. It was found that deletion of LWYIK from the gp41 fusion protein resulted in a fusion inactive virus [108]; however, this study provided evidence that this segment was needed for the enlargement of fusion pores and for post-fusion activity, rather than for interaction with cholesterol and rafts.

7   Conclusions

In this review we examine the complex structural requirements that define cholesterol-recognition motifs in membrane proteins. The initial overview section introduces the reader to the subjects of affinity, specificity and stereoselectivity of the interactions of the lipid with membrane proteins, and the implications of these properties on binding to transmembrane proteins. The next sections provide a detailed dissection of the molecular aspects currently used to identify cholesterol recognition sites in membrane proteins: CRAC, CARC, SSD and amphipathic helix motifs.

The functional implications of cholesterol-recognition motifs are covered next, using two important and paradigmatic superfamilies of membrane proteins: the G-protein coupled receptors (GPCR) and ion channels, which together represent the largest collection of membrane proteins having key roles in signal recognition and signal transduction. The possible involvement of cholesterol dysfunctional conditions in neurological disorders such as Alzheimer, Parkinson or Creutzfeldt-Jakobs diseases is also discussed. This is followed by the analysis of cholesterol recognition motifs in cholesterol trafficking from the plasmalemma to intracellular compartments and the discussion of cholesterol-recognition motifs in membrane fusion, including that of virus with eukaryotic cells, HIV fusion proteins and synaptic SNARE proteins. Without attempting to provide a comprehensive coverage of cholesterol interactions with membrane proteins, the review provides a state-of-the-art overview of the key molecular aspects of the molecular partners, i.e. cholesterol and amino acid motifs in membrane-embedded regions of membrane proteins that define the physiologically relevant crosstalk between the two. This is an ongoing and continually evolving process that in future years may lead to additional novel cholesterol binding motifs that affect protein function.

References

1. Kimura T, Jennings W, Epand RM. Roles of specific lipid species in the cell and their molecular mechanism. Prog Lipid Res. 2016;62:75–92.
2. Fantini J, Garmy N, Mahfoud R, Yahi N. Lipid rafts: structure, function and role in HIV, Alzheimer's and prion diseases. Expert Rev Mol Med. 2002;4:1–22.

3. Yang ST, Kreutzberger AJB, Lee J, Kiessling V, Tamm LK. The role of cholesterol in membrane fusion. Chem Phys Lipids. 2016;199:136–43.
4. Howe V, Sharpe LJ, Alexopoulos SJ, Kunze SV, Chua NK, Li D, Brown AJ. Cholesterol homeostasis: how do cells sense sterol excess? Chem Phys Lipids. 2016;199:170–8.
5. Jafurulla M, Chattopadhyay A. Structural stringency of cholesterol for membrane protein function utilizing stereoisomers as novel tools: a review. In: Gelissen IC, Brown AJ, editors. Cholesterol homeostasis: methods and protocols. New York, NY: Springer; 2017. p. 21–39.
6. Epand RM, Rychnovsky SD, Belani JD, Epand RF. Role of chirality in peptide-induced formation of cholesterol-rich domains. Biochem J. 2005;390:541–8.
7. Sodt AJ, Sandar ML, Gawrisch K, Pastor RW, Lyman E. The molecular structure of the liquid-ordered phase of lipid bilayers. J Am Chem Soc. 2014;136:725–32.
8. Gater DL, Saurel O, Iordanov I, Liu W, Cherezov V, Milon A. Two classes of cholesterol binding sites for the beta2AR revealed by thermostability and NMR. Biophys J. 2014;107:2305–12.
9. Rouviere E, Arnarez C, Yang L, Lyman E. Identification of two new cholesterol interaction sites on the A2A adenosine receptor. Biophys J. 2017;113:2415–24.
10. Byrne EFX, Sircar R, Miller PS, Hedger G, Luchetti G, Nachtergaele S, Tully MD, Mydock-McGrane L, Covey DF, Rambo RP, et al. Structural basis of smoothened regulation by its extracellular domains. Nature. 2016;535:517–22.
11. Posada IM, Fantini J, Contreras FX, Barrantes F, Alonso A, Goni FM. A cholesterol recognition motif in human phospholipid scramblase 1. Biophys J. 2014;107:1383–92.
12. Fantini J, Di Scala C, Baier CJ, Barrantes FJ. Molecular mechanisms of protein-cholesterol interactions in plasma membranes: functional distinction between topological (tilted) and consensus (CARC/CRAC) domains. Chem Phys Lipids. 2016;199:52–60.
13. Fantini J, Barrantes FJ. How cholesterol interacts with membrane proteins: an exploration of cholesterol-binding sites including CRAC, CARC, and tilted domains. Front Physiol. 2013;4:31.
14. Rosenhouse-Dantsker A, Noskov S, Durdagi S, Logothetis DE, Levitan I. Identification of novel cholesterol-binding regions in Kir2 channels. J Biol Chem. 2013;288:31154–64.
15. Hanson MA, Cherezov V, Griffith MT, Roth CB, Jaakola VP, Chien EY, Velasquez J, Kuhn P, Stevens RC. A specific cholesterol binding site is established by the 2.8 A structure of the human beta2-adrenergic receptor. Structure. 2008;16:897–905.
16. Fantini J, Di Scala C, Evans LS, Williamson PT, Barrantes FJ. A mirror code for protein-cholesterol interactions in the two leaflets of biological membranes. Sci Rep. 2016;6:21907.
17. Jaremko L, Jaremko M, Giller K, Becker S, Zweckstetter M. Structure of the mitochondrial translocator protein in complex with a diagnostic ligand. Science. 2014;343:1363–6.
18. Li H, Papadopoulos V. Peripheral-type benzodiazepine receptor function in cholesterol transport. Identification of a putative cholesterol recognition/interaction amino acid sequence and consensus pattern. Endocrinology. 1998;139:4991–7.
19. Baier CJ, Fantini J, Barrantes FJ. Disclosure of cholesterol recognition motifs in transmembrane domains of the human nicotinic acetylcholine receptor. Sci Rep. 2011;1:69.
20. Palmer M. Cholesterol and the activity of bacterial toxins. FEMS Microbiol Lett. 2004;238:281–9.
21. Nishio M, Umezawa Y, Fantini J, Weiss MS, Chakrabarti P. CH-pi hydrogen bonds in biological macromolecules. Phys Chem Chem Phys. 2014;16:12648–83.
22. Lee AG. Lipid-protein interactions in biological membranes: a structural perspective. Biochim Biophys Acta. 2003;1612:1–40.
23. Fantini J, Yahi N. Brain lipids in synaptic function and neurological disease. In: Clues to innovative therapeutic strategies for brain disorders. San Francisco, CA: Elsevier; 2015.
24. Fantini J, Barrantes FJ. How membrane lipids control the 3D structure and function of receptors. AIMS Biophysics. 2018;5:22–35.
25. Pydi SP, Jafurulla M, Wai L, Bhullar RP, Chelikani P, Chattopadhyay A. Cholesterol modulates bitter taste receptor function. Biochim Biophys Acta. 2016;1858:2081–7.

26. Sharpe LJ, Rao G, Jones PM, Glancey E, Aleidi SM, George AM, Brown AJ, Gelissen IC. Cholesterol sensing by the ABCG1 lipid transporter: requirement of a CRAC motif in the final transmembrane domain. Biochim Biophys Acta. 2015;1851:956–64.
27. Robinson LE, Shridar M, Smith P, Murrell-Lagnado RD. Plasma membrane cholesterol as a regulator of human and rodent P2X7 receptor activation and sensitization. J Biol Chem. 2014;289:31983–94.
28. Singh AK, McMillan J, Bukiya AN, Burton B, Parrill AL, Dopico AM. Multiple cholesterol recognition/interaction amino acid consensus (CRAC) motifs in cytosolic C tail of Slo1 subunit determine cholesterol sensitivity of Ca2+- and voltage-gated K+ (BK) channels. J Biol Chem. 2012;287:20509–21.
29. Jamin N, Neumann JM, Ostuni MA, Vu TK, Yao ZX, Murail S, Robert JC, Giatzakis C, Papadopoulos V, Lacapere JJ. Characterization of the cholesterol recognition amino acid consensus sequence of the peripheral-type benzodiazepine receptor. Mol Endocrinol. 2005;19:588–94.
30. Epand RM. Cholesterol and the interaction of proteins with membrane domains. Prog Lipid Res. 2006;45:279–94.
31. Epand RF, Thomas A, Brasseur R, Vishwanathan SA, Hunter E, Epand RM. Juxtamembrane protein segments that contribute to recruitment of cholesterol into domains. Biochemistry. 2006;45:6105–14.
32. Ulmschneider MB, Sansom MS. Amino acid distributions in integral membrane protein structures. Biochim Biophys Acta. 2001;1512:1–14.
33. Ferraro M, Masetti M, Recanatini M, Cavalli A, Bottegoni G. Mapping cholesterol interaction sites on serotonin transporter through coarse-grained molecular dynamics. PLoS One. 2016;11:e0166196.
34. Morrill GA, Kostellow AB, Gupta RK. The role of receptor topology in the vitamin D3 uptake and Ca(2+) response systems. Biochem Biophys Res Commun. 2016;477:834–40.
35. Morrill GA, Kostellow AB, Gupta RK. Computational analysis of the extracellular domain of the Ca(2)(+)-sensing receptor: an alternate model for the Ca(2)(+) sensing region. Biochem Biophys Res Commun. 2015;459:36–41.
36. Marsh D, Barrantes FJ. Immobilized lipid in acetylcholine receptor-rich membranes from Torpedo marmorata. Proc Natl Acad Sci U S A. 1978;75:4329–33.
37. Barrantes FJ. Structural basis for lipid modulation of nicotinic acetylcholine receptor function. Brain Res Brain Res Rev. 2004;47:71–95.
38. Song Y, Kenworthy AK, Sanders CR. Cholesterol as a co-solvent and a ligand for membrane proteins. Protein Sci. 2014;23:1–22.
39. Goldstein JL, DeBose-Boyd RA, Brown MS. Protein sensors for membrane sterols. Cell. 2006;124:35–46.
40. Kuwabara PE, Labouesse M. The sterol-sensing domain: multiple families, a unique role? Trends Genet. 2002;18:193–201.
41. Goldstein JL, Brown MS. Regulation of the mevalonate pathway. Nature. 1990;343:425–30.
42. Carette JE, Raaben M, Wong AC, Herbert AS, Obernosterer G, Mulherkar N, Kuehne AI, Kranzusch PJ, Griffin AM, Ruthel G, et al. Ebola virus entry requires the cholesterol transporter Niemann-Pick C1. Nature. 2011;477:340–3.
43. Cote M, Misasi J, Ren T, Bruchez A, Lee K, Filone CM, Hensley L, Li Q, Ory D, Chandran K, Cunningham J. Small molecule inhibitors reveal Niemann-Pick C1 is essential for Ebola virus infection. Nature. 2011;477:344–8.
44. Gay A, Rye D, Radhakrishnan A. Switch-like responses of two cholesterol sensors do not require protein oligomerization in membranes. Biophys J. 2015;108:1459–69.
45. Motamed M, Zhang Y, Wang ML, Seemann J, Kwon HJ, Goldstein JL, Brown MS. Identification of luminal loop 1 of Scap protein as the sterol sensor that maintains cholesterol homeostasis. J Biol Chem. 2011;286:18002–12.
46. Gao Y, Zhou Y, Goldstein JL, Brown MS, Radhakrishnan A. Cholesterol-induced conformational changes in the sterol-sensing domain of the Scap protein suggest feedback mechanism to control cholesterol synthesis. J Biol Chem. 2017;292:8729–37.

47. Adams CM, Reitz J, De Brabander JK, Feramisco JD, Li L, Brown MS, Goldstein JL. Cholesterol and 25-hydroxycholesterol inhibit activation of SREBPs by different mechanisms, both involving SCAP and Insigs. J Biol Chem. 2004;279:52772–80.
48. Infante RE, Abi-Mosleh L, Radhakrishnan A, Dale JD, Brown MS, Goldstein JL. Purified NPC1 protein. I. Binding of cholesterol and oxysterols to a 1278-amino acid membrane protein. J Biol Chem. 2008;283:1052–63.
49. Infante RE, Radhakrishnan A, Abi-Mosleh L, Kinch LN, Wang ML, Grishin NV, Goldstein JL, Brown MS. Purified NPC1 protein: II. Localization of sterol binding to a 240-amino acid soluble luminal loop. J Biol Chem. 2008;283:1064–75.
50. Li X, Lu F, Trinh MN, Schmiege P, Seemann J, Wang J, Blobel G. 3.3 A structure of Niemann-Pick C1 protein reveals insights into the function of the C-terminal luminal domain in cholesterol transport. Proc Natl Acad Sci U S A. 2017;114:9116–21.
51. Li X, Wang J, Coutavas E, Shi H, Hao Q, Blobel G. Structure of human Niemann-Pick C1 protein. Proc Natl Acad Sci U S A. 2016;113:8212–7.
52. Kwon HJ, Abi-Mosleh L, Wang ML, Deisenhofer J, Goldstein JL, Brown MS, Infante RE. Structure of N-terminal domain of NPC1 reveals distinct subdomains for binding and transfer of cholesterol. Cell. 2009;137:1213–24.
53. Gong X, Qian H, Zhou X, Wu J, Wan T, Cao P, Huang W, Zhao X, Wang X, Wang P, et al. Structural insights into the Niemann-Pick C1 (NPC1)-mediated cholesterol transfer and Ebola infection. Cell. 2016;165:1467–78.
54. Sever N, Song BL, Yabe D, Goldstein JL, Brown MS, DeBose-Boyd RA. Insig-dependent ubiquitination and degradation of mammalian 3-hydroxy-3-methylglutaryl-CoA reductase stimulated by sterols and geranylgeraniol. J Biol Chem. 2003;278:52479–90.
55. Chua NK, Howe V, Jatana N, Thukral L, Brown AJ. A conserved degron containing an amphipathic helix regulates the cholesterol-mediated turnover of human squalene monooxygenase, a rate-limiting enzyme in cholesterol synthesis. J Biol Chem. 2017;292:19959–73.
56. Albert AD, Young JE, Yeagle PL. Rhodopsin-cholesterol interactions in bovine rod outer segment disk membranes. Biochim Biophys Acta. 1996;1285:47–55.
57. Gimpl G, Burger K, Fahrenholz F. Cholesterol as modulator of receptor function. Biochemistry. 1997;36:10959–74.
58. Pucadyil TJ, Shrivastava S, Chattopadhyay A. Membrane cholesterol oxidation inhibits ligand binding function of hippocampal serotonin(1A) receptors. Biochem Biophys Res Commun. 2005;331:422–7.
59. Caffrey M. A comprehensive review of the lipid cubic phase or in meso method for crystallizing membrane and soluble proteins and complexes. Acta Crystallogr F Struct Biol Commun. 2015;71:3–18.
60. Caffrey M, Li D, Dukkipati A. Membrane protein structure determination using crystallography and lipidic mesophases: recent advances and successes. Biochemistry. 2012;51:6266–88.
61. Paila YD, Chattopadhyay A. Membrane cholesterol in the function and organization of G-protein coupled receptors. Subcell Biochem. 2010;51:439–66.
62. Hua T, Vemuri K, Nikas SP, Laprairie RB, Wu Y, Qu L, Pu M, Korde A, Jiang S, Ho JH, et al. Crystal structures of agonist-bound human cannabinoid receptor CB1. Nature. 2017;547:468–71.
63. Gimpl G. Interaction of G protein coupled receptors and cholesterol. Chem Phys Lipids. 2016;199:61–73.
64. Paila YD, Tiwari S, Chattopadhyay A. Are specific nonannular cholesterol binding sites present in G-protein coupled receptors? Biochim Biophys Acta. 2009;1788:295–302.
65. Clarke OB, Gulbis JM. Oligomerization at the membrane: potassium channel structure and function. Adv Exp Med Biol. 2012;747:122–36.
66. Levitan I, Fang Y, Rosenhouse-Dantsker A, Romanenko V. Cholesterol and ion channels. Subcell Biochem. 2010;51:509–49.
67. Rosenhouse-Dantsker A, Noskov S, Logothetis DE, Levitan I. Cholesterol sensitivity of KIR2.1 depends on functional inter-links between the N and C termini. Channels (Austin). 2013;7:303–12.

68. Barbera N, Ayee MAA, Akpa BS, Levitan I. Differential effects of sterols on ion channels: stereospecific binding vs stereospecific response. Curr Top Membr. 2017;80:25–50.
69. Addona GH, Sandermann H Jr, Kloczewiak MA, Husain SS, Miller KW. Where does cholesterol act during activation of the nicotinic acetylcholine receptor? Biochim Biophys Acta. 1998;1370:299–309.
70. Bukiya AN, Osborn CV, Kuntamallappanavar G, Toth PT, Baki L, Kowalsky G, Oh MJ, Dopico AM, Levitan I, Rosenhouse-Dantsker A. Cholesterol increases the open probability of cardiac KACh currents. Biochim Biophys Acta. 2015;1848:2406–13.
71. Di Scala C, Chahinian H, Yahi N, Garmy N, Fantini J. Interaction of Alzheimer's beta-amyloid peptides with cholesterol: mechanistic insights into amyloid pore formation. Biochemistry. 2014;53:4489–502.
72. Di Scala C, Troadec JD, Lelievre C, Garmy N, Fantini J, Chahinian H. Mechanism of cholesterol-assisted oligomeric channel formation by a short Alzheimer beta-amyloid peptide. J Neurochem. 2014;128:186–95.
73. Irvine GB, El-Agnaf OM, Shankar GM, Walsh DM. Protein aggregation in the brain: the molecular basis for Alzheimer's and Parkinson's diseases. Mol Med. 2008;14:451–64.
74. Harrison RS, Sharpe PC, Singh Y, Fairlie DP. Amyloid peptides and proteins in review. Rev Physiol Biochem Pharmacol. 2007;159:1–77.
75. Esparza TJ, Zhao H, Cirrito JR, Cairns NJ, Bateman RJ, Holtzman DM, Brody DL. Amyloid-beta oligomerization in Alzheimer dementia versus high-pathology controls. Ann Neurol. 2013;73:104–19.
76. Quist A, Doudevski I, Lin H, Azimova R, Ng D, Frangione B, Kagan B, Ghiso J, Lal R. Amyloid ion channels: a common structural link for protein-misfolding disease. Proc Natl Acad Sci U S A. 2005;102:10427–32.
77. Jang H, Connelly L, Arce FT, Ramachandran S, Lal R, Kagan BL, Nussinov R. Alzheimer's disease: which type of amyloid-preventing drug agents to employ? Phys Chem Chem Phys. 2013;15:8868–77.
78. Di Scala C, Yahi N, Boutemeur S, Flores A, Rodriguez L, Chahinian H, Fantini J. Common molecular mechanism of amyloid pore formation by Alzheimer's beta-amyloid peptide and alpha-synuclein. Sci Rep. 2016;6:28781.
79. Yahi N, Fantini J. Deciphering the glycolipid code of Alzheimer's and Parkinson's amyloid proteins allowed the creation of a universal ganglioside-binding peptide. PLoS One. 2014;9:e104751.
80. Fantini J, Carlus D, Yahi N. The fusogenic tilted peptide (67-78) of alpha-synuclein is a cholesterol binding domain. Biochim Biophys Acta. 2011;1808:2343–51.
81. Charloteaux B, Lorin A, Brasseur R, Lins L. The "Tilted Peptide Theory" links membrane insertion properties and fusogenicity of viral fusion peptides. Protein Pept Lett. 2009;16:718–25.
82. Kerr ID, Haider AJ, Gelissen IC. The ABCG family of membrane-associated transporters: you don't have to be big to be mighty. Br J Pharmacol. 2011;164:1767–79.
83. Vedhachalam C, Duong PT, Nickel M, Nguyen D, Dhanasekaran P, Saito H, Rothblat GH, Lund-Katz S, Phillips MC. Mechanism of ATP-binding cassette transporter A1-mediated cellular lipid efflux to apolipoprotein A-I and formation of high density lipoprotein particles. J Biol Chem. 2007;282:25123–30.
84. Neumann J, Rose-Sperling D, Hellmich UA. Diverse relations between ABC transporters and lipids: an overview. Biochim Biophys Acta. 2017;1859:605–18.
85. Mendez-Acevedo KM, Valdes VJ, Asanov A, Vaca L. A novel family of mammalian transmembrane proteins involved in cholesterol transport. Sci Rep. 2017;7:7450.
86. Risselada HJ. Membrane fusion stalks and lipid rafts: a love-hate relationship. Biophys J. 2017;112:2475–8.
87. Enrich C, Rentero C, Hierro A, Grewal T. Role of cholesterol in SNARE-mediated trafficking on intracellular membranes. J Cell Sci. 2015;128:1071–81.
88. Wasser CR, Ertunc M, Liu X, Kavalali ET. Cholesterol-dependent balance between evoked and spontaneous synaptic vesicle recycling. J Physiol. 2007;579:413–29.

89. Linetti A, Fratangeli A, Taverna E, Valnegri P, Francolini M, Cappello V, Matteoli M, Passafaro M, Rosa P. Cholesterol reduction impairs exocytosis of synaptic vesicles. J Cell Sci. 2010;123:595–605.

90. Hao M, Bogan JS. Cholesterol regulates glucose-stimulated insulin secretion through phosphatidylinositol 4,5-bisphosphate. J Biol Chem. 2009;284:29489–98.

91. Koseoglu S, Love SA, Haynes CL. Cholesterol effects on vesicle pools in chromaffin cells revealed by carbon-fiber microelectrode amperometry. Anal Bioanal Chem. 2011;400:2963–71.

92. Zhang J, Xue R, Ong WY, Chen P. Roles of cholesterol in vesicle fusion and motion. Biophys J. 2009;97:1371–80.

93. Churchward MA, Rogasevskaia T, Brandman DM, Khosravani H, Nava P, Atkinson JK, Coorssen JR. Specific lipids supply critical negative spontaneous curvature—an essential component of native Ca2+-triggered membrane fusion. Biophys J. 2008;94:3976–86.

94. Churchward MA, Rogasevskaia T, Hofgen J, Bau J, Coorssen JR. Cholesterol facilitates the native mechanism of Ca2+-triggered membrane fusion. J Cell Sci. 2005;118:4833–48.

95. Manes S, del Real G, Martinez AC. Pathogens: raft hijackers. Nat Rev Immunol. 2003;3:557–68.

96. Yang ST, Kiessling V, Simmons JA, White JM, Tamm LK. HIV gp41-mediated membrane fusion occurs at edges of cholesterol-rich lipid domains. Nat Chem Biol. 2015;11:424–31.

97. Scheiffele P, Roth MG, Simons K. Interaction of influenza virus haemagglutinin with sphingolipid-cholesterol membrane domains via its transmembrane domain. EMBO J. 1997;16:5501–8.

98. Freed EO. HIV-1 assembly, release and maturation. Nat Rev Microbiol. 2015;13:484–96.

99. Lee J, Nyenhuis DA, Nelson EA, Cafiso DS, White JM, Tamm LK. Structure of the Ebola virus envelope protein MPER/TM domain and its interaction with the fusion loop explains their fusion activity. Proc Natl Acad Sci U S A. 2017;114:E7987–e7996.

100. Barrett PJ, Song Y, Van Horn WD, Hustedt EJ, Schafer JM, Hadziselimovic A, Beel AJ, Sanders CR. The amyloid precursor protein has a flexible transmembrane domain and binds cholesterol. Science. 2012;336:1168–71.

101. Klug YA, Rotem E, Schwarzer R, Shai Y. Mapping out the intricate relationship of the HIV envelope protein and the membrane environment. Biochim Biophys Acta. 2017;1859:550–60.

102. Epand RM, Sayer BG, Epand RF. Peptide-induced formation of cholesterol-rich domains. Biochemistry. 2003;42:14677–89.

103. Epand RF, Sayer BG, Epand RM. The tryptophan-rich region of HIV gp41 and the promotion of cholesterol-rich domains. Biochemistry. 2005;44:5525–31.

104. Vishwanathan SA, Thomas A, Brasseur R, Epand RF, Hunter E, Epand RM. Large changes in the CRAC segment of gp41 of HIV do not destroy fusion activity if the segment interacts with cholesterol. Biochemistry. 2008;47:11869–76.

105. Vishwanathan SA, Thomas A, Brasseur R, Epand RF, Hunter E, Epand RM. Hydrophobic substitutions in the first residue of the CRAC segment of the gp41 protein of HIV. Biochemistry. 2008;47:124–30.

106. Greenwood AI, Pan J, Mills TT, Nagle JF, Epand RM, Tristram-Nagle S. CRAC motif peptide of the HIV-1 gp41 protein thins SOPC membranes and interacts with cholesterol. Biochim Biophys Acta. 2008;1778:1120–30.

107. Carravilla P, Cruz A, Martin-Ugarte I, Oar-Arteta IR, Torralba J, Apellaniz B, Perez-Gil J, Requejo-Isidro J, Huarte N, Nieva JL. Effects of HIV-1 gp41-derived virucidal peptides on virus-like lipid membranes. Biophys J. 2017;113:1301–10.

108. Chen SS, Yang P, Ke PY, Li HF, Chan WE, Chang DK, Chuang CK, Tsai Y, Huang SC. Identification of the LWYIK motif located in the human immunodeficiency virus type 1 transmembrane gp41 protein as a distinct determinant for viral infection. J Virol. 2009;83:870–83.

Crystallographic Studies of Steroid-Protein Interactions

Arthur F. Monzingo

Abstract Steroid molecules have a wide range of function in eukaryotes, including the control and maintenance of membranes, hormonal control of transcription, and intracellular signaling. X-ray crystallography has served as a successful tool for gaining understanding of the structural and mechanistic aspects of these functions by providing snapshots of steroids in complex with various types of proteins. These proteins include nuclear receptors activated by steroid hormones, several families of enzymes involved in steroid synthesis and metabolism, and proteins involved in signaling and trafficking pathways. Proteins found in some bacteria that bind and metabolize steroids have been investigated as well. A survey of the steroid-protein complexes that have been studied using crystallography and the insight learned from them is presented.

Keywords Protein-steroid complex · Ligand binding pocket · Protein structure · Nuclear receptor · Steroid metabolism · Steroid trafficking

1 Introduction

Essentially all eukaryotic cells use steroids in order to control the fluidity and flexibility of their cell membranes [1], and many cells use them as precursors of hormones and other biologically active compounds. Many different types of proteins are involved in the metabolism and other activities of steroids. Numerous crystal structures of proteins bound with steroids have been reported in the literature, with the structure coordinates deposited in the Protein Data Bank [2]. These structures have provided much insight into the function and mechanism of action of steroids in nature. The binding of steroid hormones to several families of nuclear receptors found in humans and other mammals has been investigated as well as the binding of steroids to several families of enzymes involved in steroid metabolism. These

A. F. Monzingo (✉)
Center for Biomedical Research Support, University of Texas at Austin, Austin, TX, USA
e-mail: art.monzingo@mail.utexas.edu

© Springer Nature Switzerland AG 2019   27
A. Rosenhouse-Dantsker, A. N. Bukiya (eds.), *Direct Mechanisms in Cholesterol Modulation of Protein Function*, Advances in Experimental Medicine and Biology 1135, https://doi.org/10.1007/978-3-030-14265-0_2

enzyme families include NAD(P)(H)-dependent aldo-keto reductases and short-chain dehydrogenases/reductases, as well as cytochrome P450 enzymes. The structures of proteins involved in the intracellular signaling, trafficking and regulation of steroids have also been studied. In addition, several bacterial enzymes that metabolize steroids have been investigated.

2 Cytrochrome P450 Enzymes

The cytochrome P450 enzymes comprise a superfamily of hemoproteins that participate in an array of metabolic processes. Cytochrome P450 enzymes have been identified in all kingdoms of life [3]. Members of this family are unified by a common fold and yet catalyze diverse reactions. In humans alone, there are over 50 P450s that can be divided into classes based on their intracellular localization and requirement for redox partners, which provide electrons for the monooxygenase reaction [4]. Several cholesterol-metabolizing P450s have been identified and studied. These enzymes share low amino acid sequence identity (<25%) and produce different cholesterol metabolites [5] yet bind cholesterol tightly and represent a unique system to study enzyme adaption to physiological requirements. These enzymes are characterized by a long banana-shaped steroid-binding pocket which facilitates the orientation of the steroid such that the reactive atoms are proximal to the catalytic heme iron.

In vertebrates, all steroid hormones are synthesized from pregnenolone, which is in turn formed from cholesterol via a three-step process catalyzed by cytochrome P450 11A1 (CYP11A1) [6]. During the first step, cholesterol is converted to 22R-hydroxycholesterol, the second step produces 20α,22R-dihydroxycholesterol, and the third step involves the cleavage of the C20-C22 bond in 20α,22R-dihydroxycholesterol to yield pregnenolone [7]. Sterol intermediates do not accumulate during the conversion of cholesterol to pregnenolone and bind much more tightly to CYP11A1 than cholesterol, suggesting that they remain in the active site until all three oxidative steps are completed [8]. All enzymatic steps take place in the inner mitochondrial membrane where CYP11A1 receives electrons from adrenodoxin reductase, via the [2Fe-2S] adrenodoxin.

Crystal structures of the human CYP11A1 bound with the substrate cholesterol and several reaction intermediates (including 22R-hydroxycholesterol and 20α,22R-dihydroxycholesterol) and the bovine homolog with 22R-hydroxycholesterol have provided insight into the mechanisms of the hydroxylation of cholesterol and the subsequent C-C bond cleavage [4, 9]. In each of these structures, the C20-C22 bond is nearest the heme iron. The structure of the complex of CYP11A1 bound with cholesterol and adrenodoxin is shown in Fig. 1.

The major mineralocorticoid hormone aldosterone plays a key role in the regulation of electrolyte balance and blood pressure. Excess aldosterone levels are implicated in the pathogenesis of hypertension and heart failure [10]. Cytochrome P450 11B2 (CYP11B2 or aldosterone synthase) is the sole enzyme responsible for the

Fig. 1 Complex of human CYP11A1, bound with cholesterol, and adrenodoxin (PDB ID: 3N9Y) [4]. CYP11A1 is shown as a cartoon in green, and adrenodoxin is shown in magenta. Cholesterol is shown with blue bonds, the heme as red, and the iron-sulfur cluster with cyan bonds

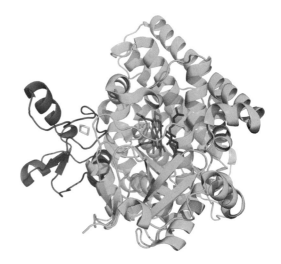

production of aldosterone in humans. CYP11B2 enzyme is localized in the adrenal cortex, where it catalyzes the conversion of deoxycorticosterone to aldosterone. The crystal structures of human CYP11B2 in complex with substrate deoxycorticosterone and an inhibitor fadrozole revealed a hydrophobic cavity with specific features associated with corticosteroid recognition [11]. In the bound deoxycorticosterone, C11 lies nearest the heme iron; C11 is hydroxylated in the conversion to aldosterone.

CYP17A1 is a membrane-bound dual-function monooxygenase with a critical role in the synthesis of many human steroid hormones [12]. The 17α-hydroxylase activity of CYP17A1 is required for the generation of glucocorticoids such as cortisol, but both the hydroxylase and 17,20-lyase activities of CYP17A1 are required for the production of androgenic and estrogenic sex steroids. CYP17A1 is an important target for the treatment of breast and prostate cancers that proliferate in response to estrogens and androgens [13]. The crystal structure of CYP17A1 bound with abiraterone, a steroid inhibitor approved as a prostate cancer drug, showed an unexpected mode of binding with the nitrogen of a pyridine ring moiety coordinating the heme iron at a distance of 2 Å [14].

Evidence indicates that neurodegeneration and development of neurological disorders are associated with disturbances in cholesterol homeostasis in the brain and that the conversion of cholesterol to 24S-hydroxycholesterol is an important mechanism that controls cholesterol turnover in the central nervous system [15]. Cholesterol 24-hydroxylation is carried out by cytochrome P450 46A1 (CYP46A1) and represents the first step in the major pathway for cholesterol elimination from the brain [16]. Unlike cholesterol, 24S-hydroxycholesterol can cross the blood–brain barrier and be delivered to the liver for further degradation. Comparison of the structures of CYP46A1 with the substrate analog cholesterol 3-sulfate and apo-CYP46A1 indicated a substantial conformational changes induced by substrate-binding, suggesting that the enzyme's flexibility makes it a potential target for therapeutic agents structurally unrelated to the natural substrate [17]. In the structure of the bound substrate analog, C24 lies closest to the heme iron with the steroid ring structure away from the heme in the extended hydrophobic pocket.

CYP51 (or lanosterol 14α-demethylase) is considered to be among the most ancient cytochrome P450 families and is found in all kingdoms from bacteria to animals [18]. CYP51 removes the 14α-methyl group from lanosterol to yield a key precursor in cholesterol and ergosterol biosynthesis. The structure of the full-length membrane monospanning CYP51 from *Saccharomyces cerevisiae* showed how the N-terminal amphipathic helix and a subsequent transmembrane helix orient the catalytic domain partly in the lipid bilayer. Bound lanosterol lies near the likely substrate and product channels connecting the active site with the lipid bilayer [19].

3   Oxysterol Regulation, Trafficking, and Signaling

Oxysterol-binding proteins are implicated in the regulation of sterol homeostasis and in signal transduction pathways in eukaryotes [20, 21]. The structures of the yeast oxysterol-binding protein Osh4 bound with several sterols, including cholesterol, revealed that the sterol molecule is bound in a hydrophobic tunnel with the 3-hydroxyl group at the bottom and the C17 side chain adjacent to a lid enclosing the tunnel [22]. Osh4 bound with cholesterol is shown in Fig. 2. The structure of Osh4 bound with 16, 22-diketocholesterol revealed what may be an intermediate conformation; that is, one between the open and closed conformations observed previously [23].

Intracellular cholesterol trafficking is central to the distribution of dietary cholesterol for utilization in membrane synthesis, synthesis of sterol hormones. Export of LDL-derived cholesterol from lysosomes requires the cooperation of the integral

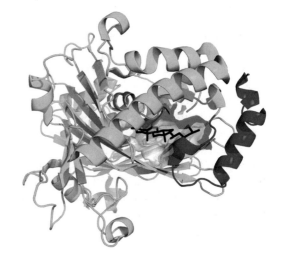

Fig. 2 Oxysterol-binding protein Osh4 bound with cholesterol (PDB ID: 1ZHY) [22]. The yeast Osh4 is shown as a cartoon in green. The N-terminal lid (residues 1–29) is shown in blue. Bound cholesterol is shown with black bonds. The oxysterol binding pocket is shown in gray

membrane protein Niemann–Pick C1 (NPC1) and a soluble protein, Niemann–Pick C2 (NPC2). Genes for these two proteins are involved in Niemann-Pick type C disease, a fatal recessive hereditary disorder characterized by the accumulation of cholesterol in lysosomes [24]. The crystal structure of NPC2 bound with cholesterol-3-O-sulfate revealed the sterol bound in a deep hydrophobic pocket sandwiched between the two β-sheets of NPC2, with only the sulfate substituent of the ligand exposed to solvent [25]. Comparison to the apo-NPC2 structure [26] indicated that binding is facilitated by a slight separation of the β-strands and substantial reorientation of side chains.

Crystal structures of the N-terminal domain of NPC1 bound with cholesterol and 25-hydroxycholesterol revealed a deep hydrophobic pocket that surrounds the sterol [27]. The 3-hydroxyl group and the tetracyclic ring are buried, but the C17 isooctyl side chain is partially exposed. This orientation is opposite to that observed with cholesterol bound to NPC2. NPC2 binds to NPC1's middle domain, and the crystal structure of that complex has been determined. Docking of the NPC1 middle domain–NPC2 complex onto the full-length NPC1 structure [28, 29] revealed a direct cholesterol transfer tunnel between NPC2 and NTD cholesterol binding pockets [30].

The binding of sterols to the membrane-bound protein Smoothened triggers Hedgehog signaling, a pathway critical in embryogenesis in vertebrates. Crystal structures of the extracellular domain of Smoothened from African clawed frog (*Xenopus laevis*) indicate a large conformational change when cholesterol is bound, as the protein goes from an open, inactive, conformation to a closed, active one [31]. Cholesterol is bound in a shallow hydrophobic pocket near the surface of the protein molecule; protein side chains hydrogen-bond with the hydroxyls of cholesterol.

4   Nuclear Steroid Receptors

There are several families of nuclear receptors which bind steroid hormones and act as ligand-activated transcription factors [32]. These families are part of the superfamily of proteins classified as *retinoid-X receptors* by the CATH Protein Structure Classification database [33]. In general, these receptors contain a C-terminal DNA-binding domain (DBD) and an N-terminal ligand binding domain (LBD). When the LBD is bound by a specific steroid hormone, there is a dimerization of the receptor; it then enters the nucleus where the DBD binds to DNA, acting as a transcription factor. Ligand-dependent activation of transcription by nuclear receptors is mediated by interactions with coactivators. Receptor agonists promote coactivator binding, and antagonists block coactivator binding [34].

The LBDs of the estrogen receptor (ER), progesterone receptor (PR), androgen receptor (AR), glutocorticoid receptor (GR), and mineralocorticoid receptor (MR) are structurally homologous, composed of 12 α-helices and a four strand β-sheet folded into a 3-layer sandwich. Steroid hormones bind within a hydrophobic pocket in the core of the LBD. The binding of the hormone causes a conformational change:

the C-terminal helix 12 moves to a new position where it can interact with coactivator proteins. Hormone antagonists bind the same site as agonists but, in general, have a bulky side chain that sticks out of the ligand-binding pocket and prevents movement of helix 12 into a position favorable for coactivator recruitment.

The binding pocket of each receptor's LBD is subtly tailored for its particular class of hormone. Structural studies, as well as experiments using site-directed mutagenesis, indicate that specific, but minor, changes of amino acids within the ligand-binding pocket can lead to dramatic changes in the hormone-binding specificity of steroid receptors [35, 36]. Based on structure-function studies and phylogenetic analyses, a series of minor amino acid changes that may account for broad changes in hormone specificity during the evolution of steroid receptors has been outlined [37]. ER has a sequence homology of only ~20% with the other nuclear steroid hormone receptors and appears to be evolutionarily the oldest. In the LBD of ER, the 3-hydroxyl oxygen of a bound estrogen is hydrogen-bonded by conserved arginine and glutamic acid residues. PR, AR, GR, and MR share a sequence homology of around 55%. With these four receptors, the 3-keto oxygen of the bound steroid is hydrogen-bonded by conserved arginine and glutamine residues. GR and MR are the most closely related with a sequence homology of 58%. Cortisol (or corticosterone in rodents) binds to and activates both the MR and GR. To prevent inappropriate MR activation by glucocorticoids in cells containing MR, 11β-hydroxysteroid dehydrogenase converts cortisol (or corticosterone) to inactive metabolites [38].

ER is activated by the binding of estrogens, including the primary estrogen hormone, estradiol [39]. Estrogens are responsible for the development and regulation of the female reproductive system and secondary sex characteristics. In ER, the 3-hydroxyl oxygen of estradiol is hydrogen-bonded by a conserved arginine and glutamic acid, rather than glutamine seen with other nuclear steroid receptors. The 17-oxygen forms a hydrogen bond with a histidine side chain. A comparison of structures of the ER LBD bound with 17β-estradiol and with antagonist raloxifene indicated a large conformational difference involving a helix (helix 12) at the C-terminus of the domain. When estradiol is bound, the helix is aligned over the ligand-binding cavity; the binding of raloxifene prevents this conformation, causing helix 12 to be aligned away from the cavity, adjacent to helix 5 [40]. This change in the orientation of helix 12 was observed in another study comparing the binding to ER with agonist diethylstilbestrol and with antagonist 4-hydroxytamoxifen. A structure of ER LBD bound with the agonist diethylstilbestrol and a co-activator peptide showed helix 12 positioned over the cavity and the co-activator adjacent to helix 5, similar to the position of helix 12 in the observed non-productive binding with antagonist [41].

PR is activated by the binding of progestins, of which progesterone is the steroid hormone required for pregnancy and the menstrual cycle. The structure of the LBD of the progesterone receptor (PR) bound with progesterone explained the receptor's selective affinity for progesterone and synthetic progestins [42]. The 3-keto oxygen of progesterone is hydrogen bonded by arginine and glutamine side chains. The shape of the hydrophobic pocket surrounding the remainder of the bound

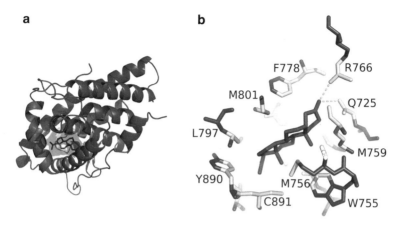

Fig. 3 Structure of the LBD of the human progesterone receptor bound with progesterone (PDB ID: 1A28) [42]. (**a**) The overall structure. A cartoon representation of the LBD of PR is shown in blue. Helix 12, which undergoes a conformational change on the binding of the progesterone, is shown in magenta. The bound steroid hormone progesterone is shown with red bonds. The hormone-binding pocket is colored gray. (**b**) Progesterone binding site. Hydrogen bonds between the ligand keto O3 atom and side chains of residues Q725 and R766 are shown with dashed lines. Hydrophobic amino acid side chains surround the steroid ring structure

progesterone contours around the methyl groups at positions 11, 18 and 19. It is unlikely that larger groups at those positions would fit in the pocket. A cartoon representation of the LBD of PR bound with progesterone is shown in Fig. 3a; progesterone in its binding site is shown in Fig. 3b.

Androgens are steroid hormones that regulate the development and maintenance of male characteristics in vertebrates. The androgen receptor (AR) is activated by either of the androgenic hormones, testosterone or dihydrotestosterone. The specific affinity of AR has been investigated with the structure of the AR LBD bound with the androgenic ligand metribolone [43, 44]. In this structure, the 3-keto oxygen of the bound ligand is hydrogen bonded by arginine and glutamine side chains and the 17-oxygen is bonded by the side chains of an asparagine and threonine. The same ligand bound to PR lacked one of the hydrogen bonds to the 17-oxygen because PR has a cysteine residue rather than the threonine of AR. Hydrogen bonds to the 17-oxygen were also observed with bound dihydrotestosterone [45]. A comparison of the structures of the AR LBD bound with testosterone, dihydrotestosterone, and the sport doping androgenic steroid tetrahydrogestrinone (THG) revealed that THG's greater affinity is due to a larger number of hydrophobic contacts [44].

Glucocorticoid steroid hormones are involved in glucose homeostasis, bone turnover, cell differentiation, and inflammation [46]. The glucocorticoid receptor (GR) is activated by the binding of cortisol or other glucocorticoids. The structures of the GR LBD bound with its endogenous ligand cortisol and with corticosteroid analogs showed the 11-oxygen and 19-oxygen of the bound ligand hydrogen bonded with asparagine and glutamine side chains [47, 48]. The greater affinity of analogs dexamethasone and mometasone furoate was attributed to the greater flexibility of the A ring of cortisol due to a rotation about the C1-C2 bond. The binding of GR with its endogenous ligand cortisol and analogs has given insight into its specific affinity [47, 48].

Mineralocorticoids are produced in the adrenal cortex and influence electrolyte balance and fluid balance [49]. Mineralocorticoid receptor (MR) is activated by the binding of mineralocorticoids and glucocorticoids. The primary mineralocorticoid in humans is aldosterone; in rats, it is corticosterone. The structures of the MR LBD bound with corticosterone revealed a hydrogen bond between the ligand 11-oxygen and an Asn residue, similar to what had been seen with GR [50, 51]. A mutation of L848 in MR to glutamine caused MR to act more like GR, with a greater affinity for cortisol. Deoxycortisone acts as an agonist with MR, even though it lacks the 11-oxygen. Progesterone is an antagonist with wild-type MR but an agonist with the S810L mutant. The crystal structure of progesterone bound with the S810L mutant showed ligand binding improved by hydrophobic contact of the leucine residue with the ligand 10-methyl moiety [51].

The mechanism of activation of the retinoic acid-related orphan nuclear receptors (ROR) is less well understood. The LBD of RORs has a sequence homology of only 11 and 22% with PR and ER, respectively, but a common fold. RORα and RORγ have been shown to bind cholesterol and hydroxycholesterol but the organization of the ligand in the binding pocket differs substantially from that observed with PR and ER [52–54].

The xenoreceptor pregnane X receptor (PXR) is a key regulator of the body's defense against foreign substances. It forms heterodimers with the retinoid X receptor and binds to responsive elements in the regulatory regions of target genes. Upon activation by xenobiotics, PXR interacts with coactivators and transcriptionally upregulates major detoxification genes such as the drug-metabolizing cytochrome P450 enzyme CYP3A4 [55]. The orientation of 17α-ethinylestradiol in the PXR binding pocket differs substantially from that observed with estrogen and ER or with cholesterol and the RORs [56]. Biophysical studies revealed that 17α-ethinylestradiol and the pesticide trans-nonachlor TNC bind cooperatively to PXR. The crystal structure showed both compounds bound in the active site.

5   Aldo-Keto Reductases

Two subfamilies of NAD(P)(H)-dependent steroid transforming aldo-keto reductase (AKR) exist in humans: hydroxysteroid dehydrogenases (HSDs; also known as AKR1C1–1C4) and 5β-reductase (AKR1D1) [57]. AKR1C enzymes catalyze the interconversion between 3-, 17-, and 20-ketosteroids and 3α/β-, 17β-, and 20α-hydroxysteroids. By functioning as HSDs, AKR1C enzymes play pivotal roles in steroid hormone action as they regulate the ratio of active and inactive androgens, estrogens, and progestogens that can bind to nuclear receptors in target tissues and hence affect transcription. They have different ratios of 3-keto-, 17-keto- and 20-ketosteroid reductase activity, and can regulate ligand access to steroid hormone

receptors in target tissues in which they are differentially expressed. AKR1D1 catalyzes the irreversible reduction of Δ^4-3-ketosteroids to 5β-dihydrosteroids to introduce a 90° bend at the steroid A/B ring junction, a characteristic structural feature
of all bile-acids, and a first step in the clearance of all steroid hormones. AKR1C
and AKR1D1 enzymes are highly homologous enzymes, sharing over 50% sequence
identity and a common (α/β)$_8$ barrel fold. Both subfamilies of enzymes bind the
steroid perpendicularly to the NAD(P)(H) cofactor in the active site so that either
the steroid A or D ring faces the nicotinamide ring. Four residues (Tyr, His, Lys,
Asp) are conserved across nearly all the AKR subfamilies and are collectively
known as the catalytic tetrad. In all AKR1D1 enzymes, the catalytic histidine is
substituted by a glutamic acid residue. In general with these enzymes, there is a
hydride transfer from the nicotinamide ring to an acceptor group, and the catalytic
tyrosine acts as the general acid/base.

   AKR1C1 is predominantly a 20α-HSD and will reduce progesterone (a potent
progestin) to 20α-hydroxyprogesterone (a weak progestin), thus regulating ligand
access to PR in target tissues [58]. AKR1C1 is also the only known NADPH-
dependent 3-ketosteroid reductase that will act as a 3β-HSD; it will convert
5α-dihydrotestosterone (5α-DHT) exclusively to 3β-androstanediol, a natural proapoptotic ligand for ERβ in target tissues [59]. Remarkably, AKR1C1 shares 98%
sequence homology but very different activity with AKR1C2, which is predominately a 3α-HSD. Human AKR1C1 was co-crystallized with substrate progesterone
and co-factor NADPH; and in the crystal structure, product 20α-hydroxyprogesterone
was found bound in the active site [60]. The overall structure of human AKRC1
bound with 20α-hydroxyprogesterone and NADP$^+$ is shown in Fig. 4a. In the

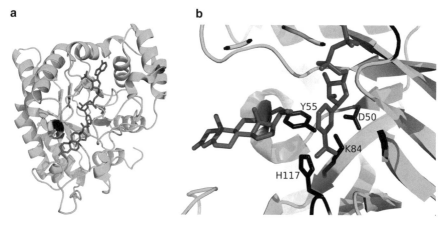

Fig. 4 Ternary complex of the human enzyme AKR1C1 bound with product
20α-hydroxyprogesterone and co-factor NADP$^+$ (PDB ID: 1MRQ) [60]. (**a**) Overall structure. The
protein is shown as a cartoon in green. Bound 20α-hydroxyprogesterone is shown with red bonds
and NADP$^+$ with blue. (**b**) Enzyme active site. Bound product 20α-hydroxyprogesterone is shown
with red bonds and co-factor NADP$^+$ with blue. Side chains of the conserved catalytic tetrad (D50,
Y55, K84, H117) are shown in black

structure, the 20α-hydroxyl is oriented near the nicotinamide ring of NADP+ and is stabilized by a hydrogen bond with the side chain of a histidine residue not conserved in AKR1C2. The binding of 20α-hydroxyprogesterone in the active site of AKRC1 is highlighted in Fig. 4b.

AKR1C2, predominately a 3α-HSD, will reduce 5α-dihydrotestosterone (a potent androgen) to 3α-androstanediol (a weak androgen), regulating ligand access to the androgen receptor (AR) in target tissues [61]. AKR1C2 also exhibits some 17β-HSD activity and will reduce androstendione to testosterone. Crystal structures of AKR1C2 have investigated both its 3α-HSD and its 17β-HSD activities. They show that this enzyme's steroid-binding pocket possesses considerable flexibility. The human AKR1C2 (3α-HSD type 3) was co-crystallized with 5α-dihydrotestosterone in the presence of NADP+. Two products and modes of binding were observed in the crystal: epiandrosterone with the 3α-hydroxyl pointed toward the nicotinamide ring and 5α-androstane-3,17-dione with the C17 keto oriented toward the nicotamide ring [62]. To investigate its 17β-HSD activity, the human AKR1C2 was co-crystallized with testosterone and NADP+. Androstendione was found bound in the active site with the C3 keto oriented toward the nicotinamide ring, the implication being that the oxidation of testosterone to androstendione had occurred and then androstendione had rebound in the active site in the reverse orientation. In a prior study using crystals grown in 0.26 M acetate, product testosterone was found bound in the active site oriented with the C17 toward the NADP+ and with an acetate molecule lying adjacent to the nicotinamide ring of NADP+ [63]. As described above, AKR1C2 shares 98% sequence homology but different activity with AKR1C1, which is predominately a 20α-HSD. In co-crystals of AKR1C2 with progesterone and NADP+, the progesterone is observed with two different bonding modes, both with C20 oriented toward the nicotinamide. With co-crystals of the V74L variant with progesterone and NADP+, progesterone adopts an orientation similar to that observed with 20-hydroxyprogesterone and AKR1C1. 20α-HSD activity in the variant is also enhanced [64].

AKR1C3 is predominately a 17β-HSD and will reduce Δ_4-androstene-3,17-dione (weak androgen) to testosterone (potent androgen), regulating access of testosterone to the AR in target tissues. AKR1C3 also reduces estrone (weak estrogen) to 17β-estradiol (potent estrogen), regulating ligand access to ER in target tissues. Ternary complexes of human AKR1C3 showed different binding modes for substrate androstenedione and product testosterone in the active site [65].

The structure of the ternary complex of the rat AKR1C9 (3α-HSD), NADP+, and the competitive inhibitor testosterone showed the C3 ketone, corresponding to the reactive group in a substrate, poised above the nicotinamide ring which is involved in hydride transfer. In addition, the C3 ketone forms hydrogen bonds with the Tyr and His active-site residues [66]. The structure of the ternary complex of the rabbit AHR1C5 (17α-HSD) with testosterone and NADP also showed the C3 ketone oriented toward the nicotinamide ring [67]. In the structure of the ternary complex of the mouse AKR1C21 (17α-HSD) with product epi-testosterone and NADP, the 17-hydroxyl of epi-testosterone is oriented near the nicotinamide [68].

AKR1D1 (Δ^4-3-ketosteroid 5β-reductase) is a steroid double-bond reductase. This liver-specific enzyme converts any Δ^4-3-ketosteroid into a 5β-dihyrosteroid, converting the A/B ring juncture in the steroid to a cis-configuration. This introduces a 90° bend in the steroid structure, which is a characteristic structural feature of all bile-acids [57]. Ternary complexes of the enzyme and NADP+ with progesterone and with cortisone showed the ligand bound with carbon-carbon double bond in proximity to the nicotinamide ring of NADP+ [69]. The structure of the ternary complex with NADP+ and the product 5β-dihydroprogesterone revealed the active site Glu residue not interacting with other catalytic residues, suggesting that it may be involved in substrate binding but not catalysis [70]. Mutation of the active site Glu to His converts the enzyme to a 3β-HSD; and structures with epiandrosterone, 5β-dihydrotestosterone, and Δ^4-androstene-3,17-dione provided insight into the structural basis for this change [71].

6 Short-Chain Dehydrogenases/Reductases

Short-chain dehydrogenases/reductases (SDRs) constitute a large family of NAD(P) (H)-dependent oxidoreductases [72, 73]. Despite low sequence identities between different forms, the 3D structures display highly similar α/β folding patterns with a central β-sheet typical of the Rossmann-fold. The conservation of this motif for nucleotide binding and an active site provide a platform for various enzymatic activities. As with the NAD(P)(H)-dependent aldo-keto reductase family, there is a hydride transfer from the nicotinamide ring to an acceptor group, and the catalytic tyrosine acts as the general acid/base. At least two SDR enzymes play critical roles in steroid hormone metabolism.

Human 17β-hydroxysteroid dehydrogenase (17β-HSD) is an SDR enzyme that plays a major role in the formation of active estrogens in gonadal and peripheral tissues. 17β-HSD catalyzes the reversible conversion of the biologically inactive estrogen, estrone, to 17β-estradiol, the most active estrogen. This enzyme has been demonstrated to be involved in maintaining high 17β-estradiol levels in breast tumors of postmenopausal women [74]. Structures of 17β-HSD bound with 17β-estradiol, both with and without NADP+, reveal hydrogen-bonding contacts of the substrate with SDR-conserved active site Lys and Tyr residues and a histidine residue [75–77]. The ternary complex of 17β-HSD bound with 17β-estradiol and NADP+ is shown in Fig. 5a. The steroid is stabilized by hydrophobic residues in the binding pocket. In the ternary complex, the 17β hydroxyl lies near the nicotinamide ring. The binding of 17β-estradiol bound in the active site of 17β-HSD is shown in Fig. 5b. Comparison of the structures of several complexes of 17β-HSD and non-estrogen steroids with the 17β-estradiol complex showed that an active site leucine residue contributed to the enzyme's preference for estrogens. The side chain of L149 prevents steroids with a group in the C19 position from binding in an orientation productive for catalysis [78, 79].

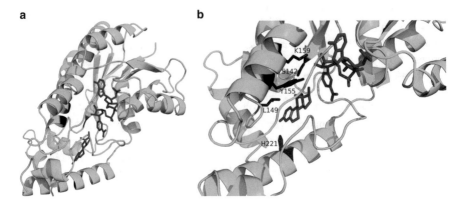

Fig. 5 Ternary complex of the human enzyme 17β-HSD bound with product 17β-estradiol and co-factor NADP+ (PDB ID: 1FDT) [76]. (**a**) Overall structure. The protein is shown as a cartoon in green. Bound 17β-estradiol is shown with red bonds and NADP+ with blue. (**b**) Enzyme active site. Bound product 17β-estradiol is shown with red bonds and co-factor NADP+ with blue. Side chains of the conserved catalytic triad (S142, Y155, K159) are shown in black. The side chain of a conserved histidine (H221), which is believed to also assist in catalysis, is also shown in black. The side chain of L149, which contributes to the enzyme's preference for estrogens, is shown in black

Glucocorticoids play important roles in a variety of physiological and cellular processes [80]. The active hydroxyl form of glucocorticoids (such as cortisol) binds to nuclear receptors and subsequently influences gene transcription. There are two isoforms of 11β-hydroxysteroid dehydrogenase (11β-HSD). 11β-HSD type1 is an NADPH-dependent reductase that converts the inactive 11β glucocorticoids to their active form. 11β-HSD type 2 acts as a dehydrogenase catalyzing the conversion of cortisol to cortisone. A level of cortisol regulation is achieved by the differential tissue distribution of the two isoforms. In the structure of murine 11β-HSD type 1 bound with NADPH and corticosterone, the 11β-hydroxyl of the substrate is near the nicotinamide ring [81].

7 Bacterial Enzymes

The limited role of steroids in bacteria is not well understood although it is known that some bacteria use sterols to build their cell envelope [82]. Sterol-producing cytochrome P450s in bacteria are generally soluble, not membrane-bound; and there is keen interest in the industrial application of these enzymes in the production of steroid hormones [83]. To date, there have been several crystallographic studies of bacterial P450s bound with steroids. The structure of the soluble CYP51 from *Mycobacterium tuberculosis* bound with substrate analog estriol indicated that substrate binding causes conformational changes in the protein [84]. P450eryF (CYP107A family) from *Saccharopolyspora erythraea* was found to have two molecules of androstenedione bound in the active site, suggesting a mechanism for the homotropic cooperativity observed with some eukaryotic P450s [85]. The structures of CYP154C5 from *Nocardia farcinica* bound with four different steroid molecules gave insight into the stereoselectivity in the conversion of pregnans and androstans to 16α-hydroxylated steroid products [86]. Enzymatic studies showed that CYP109E1 from *Bacillus megaterium* DSM319 can hydroxylate testosterone at position 16β. They also showed that corticosterone, with a bulky substituent attached at the 17 position, binds the enzyme but is not converted. A crystal structure showed testosterone bound in a nonproductive mode with the C3 keto oxygen coordinating the heme iron. The structure of CYP109E1 with corticosterone bound showed a large conformational change in the active site with four corticosterone molecules bound [87].

The SDR 3α,20β-HSD from *Streptomyces hydrogenans* reversibly oxidizes the 3α-hydroxyl and 20β-hydroxyl groups of androstane and pregnane derivatives in a reaction requiring NAD(H) [88]. The ternary structure of 3α,20β-HSD with a steroid inhibitor bound helped identify a catalytic triad of Tyr, Lys, and Ser residues [89]. The structure of another bacterial SDR, a steroid-inducible 3β/17β-HSD from *Comamonas testosteroni*, has also been reported [90].

3-Ketosteroid Δ^1-dehydrogenase is an FAD-dependent enzyme that catalyzes the introduction of a double bond into the C1-C2 position of the 3-ketosteroid ring. The enzyme can dehydrogenate a wide variety of 3-ketosteroids, but not 3-hydroxysteroids, and has a preference for substrates unsaturated at the C4-C5 position, such as 4-androstene-3,17-dione and cortisone. There are several FAD-dependent enzymes in the pathway of steroid ring system degradation found in many steroid-degrading bacteria [91]. To date, structures of three of these enzymes have been determined with a steroid molecule bound. 3-Ketosteroid Δ^1-dehydrogenase catalyzes the introduction of a double bond into the C1-C2 position of the 3-ketosteroid ring. The structure of the homolog from *Rhodococcus erythopolis* bound with product 1,4-androstadiene-3,17-dione has been reported [92]. 3-Ketosteroid Δ^4-(5α)-dehydrogenases introduce a double bond between the C4 and C5 atoms of 3-keto-(5α)-steroids. The structure of the *Rhodococcus jostii* homolog bound with 4-androstene-3,17-dione has been solved [93]. Cholesterol oxidase from *Brevibacterium sterolicum* catalyzes the oxidation and isomerization of 3β-hydroxy steroids with a double bond at Δ^5-Δ^6 of the steroid ring backbone. Its structure with substrate dehydroisoandrosterone has been reported [94]. These three enzymes share a common fold and are members of the *steroid monooxygenase* superfamily, although sequence homology (12–28%) is low. They contain a catalytic domain and an FAD-binding domain with the active site. In all three cases, the bound steroid lies in the active site near the flavin ring.

8   Summary and Conclusions

Steroid molecules are found throughout nature and have a wide range of function. In eukaryotes, these functions include the control and maintenance of membranes, hormonal control of transcription, and intracellular signaling. Steroids are also found in bacteria, although their role there is not well understood. The study of the interactions of steroid molecules with proteins is key to understanding the functionality and metabolism of steroids. We have presented an overview of the major classes of proteins that bind steroids and the insights that have been gained from their study using X-ray crystallography. These classes of proteins include three enzyme families that play roles in the synthesis and modification of steroids: cytochrome P450 enzymes, NAD(P)(H)-dependent aldo-keto reductases, and NAD(P)(H)-dependent SDRs. Other classes of proteins that have been studied include ones involved in intracellular trafficking and signal pathways, and the nuclear steroid hormone receptors. In addition, an FAD-dependent steroid monooxygenase family is found in bacteria.

To date, X-ray crystallography has provided deep mechanistic insight into the function of steroids by examining steroid-protein interactions at atomic resolution. This field will, no doubt, provide many intriguing findings in the future.

References

1. Ourisson G, Nakatani Y. The terpenoid theory of the origin of cellular life: the evolution of terpenoids to cholesterol. Chem Biol. 1994;1:11–23.
2. Berman HM, Westbrook J, Feng Z, Gilliland G, Bhat TN, Weissig H, et al. The protein data bank. Nucleic Acids Res. 2000;28:235–42.
3. Lamb DC, Lei L, Warrilow AG, Lepesheva GI, Mullins JG, Waterman MR, et al. The first virally encoded cytochrome p450. J Virol. 2009;83:8266–9.
4. Strushkevich N, MacKenzie F, Cherkesova T, Grabovec I, Usanov S, Park HW. Structural basis for pregnenolone biosynthesis by the mitochondrial monooxygenase system. Proc Natl Acad Sci U S A. 2011;108:10139–43.
5. Pikuleva IA. Cholesterol-metabolizing cytochromes P450. Drug Metab Dispos. 2006;34:513–20.
6. Simpson ER, Boyd GS. The cholesterol side-chain cleavage system of bovine adrenal cortex. Eur J Biochem. 1967;2:275–85.
7. Burstein S, Middleditch BS, Gut M. Mass spectrometric study of the enzymatic conversion of cholesterol to (22R)-22-hydroxycholesterol, (20R,22R)-20,22-dihydroxycholesterol, and pregnenolone, and of (22R)-22-hydroxycholesterol to the lgycol and pregnenolone in bovine adrenocortical preparations. Mode of oxygen incorporation. J Biol Chem. 1975;250:9028–37.
8. Lambeth JD, Kitchen SE, Farooqui AA, Tuckey R, Kamin H. Cytochrome P-450scc-substrate interactions. Studies of binding and catalytic activity using hydroxycholesterols. J Biol Chem. 1982;257:1876–84.
9. Mast N, Annalora AJ, Lodowski DT, Palczewski K, Stout CD, Pikuleva IA. Structural basis for three-step sequential catalysis by the cholesterol side chain cleavage enzyme CYP11A1. J Biol Chem. 2011;286:5607–13.

10. Tomaschitz A, Pilz S, Ritz E, Obermayer-Pietsch B, Pieber TR. Aldosterone and arterial hypertension. Nat Rev Endocrinol. 2010;6:83–93.
11. Strushkevich N, Gilep AA, Shen L, Arrowsmith CH, Edwards AM, Usanov SA, et al. Structural insights into aldosterone synthase substrate specificity and targeted inhibition. Mol Endocrinol. 2013;27:315–24.
12. Miller WL, Auchus RJ. The molecular biology, biochemistry, and physiology of human steroidogenesis and its disorders. Endocr Rev. 2011;32:81–151.
13. Yap TA, Carden CP, Attard G, de Bono JS. Targeting CYP17: established and novel approaches in prostate cancer. Curr Opin Pharmacol. 2008;8:449–57.
14. DeVore NM, Scott EE. Structures of cytochrome P450 17A1 with prostate cancer drugs abiraterone and TOK-001. Nature. 2012;482:116–9.
15. Björkhem I, Meaney S. Brain cholesterol: long secret life behind a barrier. Arterioscler Thromb Vasc Biol. 2004;24:806–15.
16. Björkhem I. Crossing the barrier: oxysterols as cholesterol transporters and metabolic modulators in the brain. J Intern Med. 2006;260:493–508.
17. Mast N, White MA, Bjorkhem I, Johnson EF, Stout CD, Pikuleva IA. Crystal structures of substrate-bound and substrate-free cytochrome P450 46A1, the principal cholesterol hydroxylase in the brain. Proc Natl Acad Sci U S A. 2008;105:9546–51.
18. Aoyama Y, Noshiro M, Gotoh O, Imaoka S, Funae Y, Kurosawa N, et al. Sterol 14-demethylase P450 (P45014DM*) is one of the most ancient and conserved P450 species. J Biochem. 1996;119:926–33.
19. Monk BC, Tomasiak TM, Keniya MV, Huschmann FU, Tyndall JD, O'Connell JD, et al. Architecture of a single membrane spanning cytochrome P450 suggests constraints that orient the catalytic domain relative to a bilayer. Proc Natl Acad Sci U S A. 2014;111:3865–70.
20. Beh CT, Rine J. A role for yeast oxysterol-binding protein homologs in endocytosis and in the maintenance of intracellular sterol-lipid distribution. J Cell Sci. 2004;117:2983–96.
21. Wang PY, Weng J, Anderson RG. OSBP is a cholesterol-regulated scaffolding protein in control of ERK 1/2 activation. Science. 2005;307:1472–6.
22. Im YJ, Raychaudhuri S, Prinz WA, Hurley JH. Structural mechanism for sterol sensing and transport by OSBP-related proteins. Nature. 2005;437:154–8.
23. Koag MC, Cheun Y, Kou Y, Ouzon-Shubeita H, Min K, Monzingo AF, et al. Synthesis and structure of 16,22-diketocholesterol bound to oxysterol-binding protein Osh4. Steroids. 2013;78:938–44.
24. Pentchev PG. Niemann-Pick C research from mouse to gene. Biochim Biophys Acta. 2004;1685:3–7.
25. Xu S, Benoff B, Liou HL, Lobel P, Stock AM. Structural basis of sterol binding by NPC2, a lysosomal protein deficient in Niemann-Pick type C2 disease. J Biol Chem. 2007;282:23525–31.
26. Friedland N, Liou HL, Lobel P, Stock AM. Structure of a cholesterol-binding protein deficient in Niemann-Pick type C2 disease. Proc Natl Acad Sci U S A. 2003;100:2512–7.
27. Kwon HJ, Abi-Mosleh L, Wang ML, Deisenhofer J, Goldstein JL, Brown MS, et al. Structure of N-terminal domain of NPC1 reveals distinct subdomains for binding and transfer of cholesterol. Cell. 2009;137:1213–24.
28. Li X, Wang J, Coutavas E, Shi H, Hao Q, Blobel G. Structure of human Niemann-Pick C1 protein. Proc Natl Acad Sci U S A. 2016;113:8212–7.
29. Gong X, Qian H, Zhou X, Wu J, Wan T, Cao P, et al. Structural insights into the Niemann-Pick C1 (NPC1)-mediated cholesterol transfer and Ebola infection. Cell. 2016;165:1467–78.
30. Li X, Saha P, Li J, Blobel G, Pfeffer SR. Clues to the mechanism of cholesterol transfer from the structure of NPC1 middle lumenal domain bound to NPC2. Proc Natl Acad Sci U S A. 2016;113:10079–84.
31. Huang P, Nedelcu D, Watanabe M, Jao C, Kim Y, Liu J, et al. Cellular cholesterol directly activates smoothened in hedgehog signaling. Cell. 2016;166:1176–87.e14.
32. Busillo JM, Rhen T, Cidlowski JA. Steroid hormone action. In: Strauss JF, editor. Yen and Jaffe's reproductive endocrinology. 7th ed. Philadelphia, PA: Elsevier; 2014. p. 93–107.

33. Dawson NL, Lewis TE, Das S, Lees JG, Lee D, Ashford P, et al. CATH: an expanded resource to predict protein function through structure and sequence. Nucleic Acids Res. 2017;45:D289–D95.
34. Beato M, Sánchez-Pacheco A. Interaction of steroid hormone receptors with the transcription initiation complex. Endocr Rev. 1996;17:587–609.
35. Tanenbaum DM, Wang Y, Williams SP, Sigler PB. Crystallographic comparison of the estrogen and progesterone receptor's ligand binding domains. Proc Natl Acad Sci U S A. 1998;95:5998–6003.
36. Veldscholte J, Voorhorst-Ogink MM, Bolt-de Vries J, van Rooij HC, Trapman J, Mulder E. Unusual specificity of the androgen receptor in the human prostate tumor cell line LNCaP: high affinity for progestagenic and estrogenic steroids. Biochim Biophys Acta. 1990;1052:187–94.
37. Thornton JW. Evolution of vertebrate steroid receptors from an ancestral estrogen receptor by ligand exploitation and serial genome expansions. Proc Natl Acad Sci U S A. 2001;98:5671–6.
38. Seckl JR. 11beta-hydroxysteroid dehydrogenases: changing glucocorticoid action. Curr Opin Pharmacol. 2004;4:597–602.
39. Tsai MJ, O'Malley BW. Molecular mechanisms of action of steroid/thyroid receptor super-family members. Annu Rev Biochem. 1994;63:451–86.
40. Brzozowski AM, Pike AC, Dauter Z, Hubbard RE, Bonn T, Engström O, et al. Molecular basis of agonism and antagonism in the oestrogen receptor. Nature. 1997;389:753–8.
41. Shiau AK, Barstad D, Loria PM, Cheng L, Kushner PJ, Agard DA, et al. The structural basis of estrogen receptor/coactivator recognition and the antagonism of this interaction by tamoxifen. Cell. 1998;95:927–37.
42. Williams SP, Sigler PB. Atomic structure of progesterone complexed with its receptor. Nature. 1998;393:392–6.
43. Matias PM, Donner P, Coelho R, Thomaz M, Peixoto C, Macedo S, et al. Structural evidence for ligand specificity in the binding domain of the human androgen receptor. Implications for pathogenic gene mutations. J Biol Chem. 2000;275:26164–71.
44. Pereira de Jésus-Tran K, Côté PL, Cantin L, Blanchet J, Labrie F, Breton R. Comparison of crystal structures of human androgen receptor ligand-binding domain complexed with vari-ous agonists reveals molecular determinants responsible for binding affinity. Protein Sci. 2006;15:987–99.
45. Sack JS, Kish KF, Wang C, Attar RM, Kiefer SE, An Y, et al. Crystallographic structures of the ligand-binding domains of the androgen receptor and its T877A mutant complexed with the natural agonist dihydrotestosterone. Proc Natl Acad Sci U S A. 2001;98:4904–9.
46. Reichardt HM, Tronche F, Berger S, Kellendonk C, Schütz G. New insights into glucocorticoid and mineralocorticoid signaling: lessons from gene targeting. Adv Pharmacol. 2000;47:1–21.
47. Bledsoe RK, Montana VG, Stanley TB, Delves CJ, Apolito CJ, McKee DD, et al. Crystal structure of the glucocorticoid receptor ligand binding domain reveals a novel mode of recep-tor dimerization and coactivator recognition. Cell. 2002;110:93–105.
48. He Y, Yi W, Suino-Powell K, Zhou XE, Tolbert WD, Tang X, et al. Structures and mechanism for the design of highly potent glucocorticoids. Cell Res. 2014;24:713–26.
49. Funder JW. Glucocorticoid and mineralocorticoid receptors: biology and clinical relevance. Annu Rev Med. 1997;48:231–40.
50. Li Y, Suino K, Daugherty J, Xu HE. Structural and biochemical mechanisms for the specific-ity of hormone binding and coactivator assembly by mineralocorticoid receptor. Mol Cell. 2005;19:367–80.
51. Fagart J, Huyet J, Pinon GM, Rochel M, Mayer C, Rafestin-Oblin ME. Crystal structure of a mutant mineralocorticoid receptor responsible for hypertension. Nat Struct Mol Biol. 2005;12:554–5.
52. Kallen JA, Schlaeppi JM, Bitsch F, Geisse S, Geiser M, Delhon I, et al. X-ray structure of the hRORalpha LBD at 1.63 A: structural and functional data that cholesterol or a cholesterol derivative is the natural ligand of RORalpha. Structure. 2002;10:1697–707.

53. Kallen J, Schlaeppi JM, Bitsch F, Delhon I, Fournier B. Crystal structure of the human RORalpha ligand binding domain in complex with cholesterol sulfate at 2.2 A. J Biol Chem. 2004;279:14033–8.

54. Jin L, Martynowski D, Zheng S, Wada T, Xie W, Li Y. Structural basis for hydroxycholesterols as natural ligands of orphan nuclear receptor RORgamma. Mol Endocrinol. 2010;24:923–9.

55. Lehmann JM, McKee DD, Watson MA, Willson TM, Moore JT, Kliewer SA. The human orphan nuclear receptor PXR is activated by compounds that regulate CYP3A4 gene expression and cause drug interactions. J Clin Invest. 1998;102:1016–23.

56. Delfosse V, Dendele B, Huet T, Grimaldi M, Boulahtouf A, Gerbal-Chaloin S, et al. Synergistic activation of human pregnane X receptor by binary cocktails of pharmaceutical and environmental compounds. Nat Commun. 2015;6:8089.

57. Penning TM, Drury JE. Human aldo-keto reductases: function, gene regulation, and single nucleotide polymorphisms. Arch Biochem Biophys. 2007;464:241–50.

58. Rizner TL, Smuc T, Rupreht R, Sinkovec J, Penning TM. AKR1C1 and AKR1C3 may determine progesterone and estrogen ratios in endometrial cancer. Mol Cell Endocrinol. 2006;248:126–35.

59. Steckelbroeck S, Jin Y, Gopishetty S, Oyesanmi B, Penning TM. Human cytosolic 3alpha-hydroxysteroid dehydrogenases of the aldo-keto reductase superfamily display significant 3beta-hydroxysteroid dehydrogenase activity: implications for steroid hormone metabolism and action. J Biol Chem. 2004;279:10784–95.

60. Couture JF, Legrand P, Cantin L, Luu-The V, Labrie F, Breton R. Human 20alpha-hydroxysteroid dehydrogenase: crystallographic and site-directed mutagenesis studies lead to the identification of an alternative binding site for C21-steroids. J Mol Biol. 2003;331:593–604.

61. Rizner TL, Lin HK, Peehl DM, Steckelbroeck S, Bauman DR, Penning TM. Human type 3 3alpha-hydroxysteroid dehydrogenase (aldo-keto reductase 1C2) and androgen metabolism in prostate cells. Endocrinology. 2003;144:2922–32.

62. Zhang B, Hu XJ, Wang XQ, Thériault JF, Zhu DW, Shang P, et al. Human 3α-hydroxysteroid dehydrogenase type 3: structural clues of 5α-DHT reverse binding and enzyme down-regulation decreasing MCF7 cell growth. Biochem J. 2016;473:1037–46.

63. Nahoum V, Gangloff A, Legrand P, Zhu DW, Cantin L, Zhorov BS, et al. Structure of the human 3alpha-hydroxysteroid dehydrogenase type 3 in complex with testosterone and NADP at 1.25-A resolution. J Biol Chem. 2001;276:42091–8.

64. Zhang B, Zhu DW, Hu XJ, Zhou M, Shang P, Lin SX. Human 3-alpha hydroxysteroid dehydrogenase type 3 (3α-HSD3): the V54L mutation restricting the steroid alternative binding and enhancing the 20α-HSD activity. J Steroid Biochem Mol Biol. 2014;141:135–43.

65. Qiu W, Zhou M, Labrie F, Lin SX. Crystal structures of the multispecific 17beta-hydroxysteroid dehydrogenase type 5: critical androgen regulation in human peripheral tissues. Mol Endocrinol. 2004;18:1798–807.

66. Bennett MJ, Albert RH, Jez JM, Ma H, Penning TM, Lewis M. Steroid recognition and regulation of hormone action: crystal structure of testosterone and NADP+ bound to 3 alpha-hydroxysteroid/dihydrodiol dehydrogenase. Structure. 1997;5:799–812.

67. Couture JF, Legrand P, Cantin L, Labrie F, Luu-The V, Breton R. Loop relaxation, a mechanism that explains the reduced specificity of rabbit 20alpha-hydroxysteroid dehydrogenase, a member of the aldo-keto reductase superfamily. J Mol Biol. 2004;339:89–102.

68. Faucher F, Cantin L, Pereira de Jésus-Tran K, Lemieux M, Luu-The V, Labrie F, et al. Mouse 17alpha-hydroxysteroid dehydrogenase (AKR1C21) binds steroids differently from other aldo-keto reductases: identification and characterization of amino acid residues critical for substrate binding. J Mol Biol. 2007;369:525–40.

69. Di Costanzo L, Drury JE, Penning TM, Christianson DW. Crystal structure of human liver Delta4-3-ketosteroid 5beta-reductase (AKR1D1) and implications for substrate binding and catalysis. J Biol Chem. 2008;283:16830–9.

70. Faucher F, Cantin L, Luu-The V, Labrie F, Breton R. Crystal structures of human Delta4-3-ketosteroid 5beta-reductase (AKR1D1) reveal the presence of an alternative binding site responsible for substrate inhibition. Biochemistry. 2008;47:13537–46.

71. Chen M, Drury JE, Christianson DW, Penning TM. Conversion of human steroid 5β-reductase (AKR1D1) into 3β-hydroxysteroid dehydrogenase by single point mutation E120H: example of perfect enzyme engineering. J Biol Chem. 2012;287:16609–22.

72. Oppermann U, Filling C, Hult M, Shafqat N, Wu X, Lindh M, et al. Short-chain dehydrogenases/reductases (SDR): the 2002 update. Chem Biol Interact. 2003;143-144:247–53.

73. Kavanagh KL, Jörnvall H, Persson B, Oppermann U. Medium- and short-chain dehydrogenase/reductase gene and protein families: the SDR superfamily: functional and structural diversity within a family of metabolic and regulatory enzymes. Cell Mol Life Sci. 2008;65:3895–906.

74. Miyoshi Y, Ando A, Shiba E, Taguchi T, Tamaki Y, Noguchi S. Involvement of up-regulation of 17beta-hydroxysteroid dehydrogenase type 1 in maintenance of intratumoral high estradiol levels in postmenopausal breast cancers. Int J Cancer. 2001;94:685–9.

75. Azzi A, Rehse PH, Zhu DW, Campbell RL, Labrie F, Lin SX. Crystal structure of human estrogenic 17 beta-hydroxysteroid dehydrogenase complexed with 17 beta-estradiol. Nat Struct Biol. 1996;3:665–8.

76. Breton R, Housset D, Mazza C, Fontecilla-Camps JC. The structure of a complex of human 17beta-hydroxysteroid dehydrogenase with estradiol and NADP+ identifies two principal targets for the design of inhibitors. Structure. 1996;4:905–15.

77. Mazza C, Breton R, Housset D, Fontecilla-Camps JC. Unusual charge stabilization of NADP+ in 17beta-hydroxysteroid dehydrogenase. J Biol Chem. 1998;273:8145–52.

78. Han Q, Campbell RL, Gangloff A, Huang YW, Lin SX. Dehydroepiandrosterone and dihydrotestosterone recognition by human estrogenic 17beta-hydroxysteroid dehydrogenase. C-18/c-19 steroid discrimination and enzyme-induced strain. J Biol Chem. 2000;275:1105–11.

79. Shi R, Lin SX. Cofactor hydrogen bonding onto the protein main chain is conserved in the short chain dehydrogenase/reductase family and contributes to nicotinamide orientation. J Biol Chem. 2004;279:16778–85.

80. Stewart PM, Krozowski ZS. 11 beta-hydroxysteroid dehydrogenase. Vitam Horm. 1999;57:249–324.

81. Zhang J, Osslund TD, Plant MH, Clogston CL, Nybo RE, Xiong F, et al. Crystal structure of murine 11 beta-hydroxysteroid dehydrogenase 1: an important therapeutic target for diabetes. Biochemistry. 2005;44:6948–57.

82. Brennan PJ, Nikaido H. The envelope of mycobacteria. Annu Rev Biochem. 1995;64:29–63.

83. Bernhardt R. Cytochromes P450 as versatile biocatalysts. J Biotechnol. 2006;124:128–45.

84. Podust LM, Yermalitskaya LV, Lepesheva GI, Podust VN, Dalmasso EA, Waterman MR. Estriol bound and ligand-free structures of sterol 14alpha-demethylase. Structure. 2004;12:1937–45.

85. Cupp-Vickery J, Anderson R, Hatziris Z. Crystal structures of ligand complexes of P450eryF exhibiting homotropic cooperativity. Proc Natl Acad Sci U S A. 2000;97:3050–5.

86. Herzog K, Bracco P, Onoda A, Hayashi T, Hoffmann K, Schallmey A. Enzyme-substrate complex structures of CYP154C5 shed light on its mode of highly selective steroid hydroxylation. Acta Crystallogr D Biol Crystallogr. 2014;70:2875–89.

87. Jóźwik IK, Kiss FM, Gricman Ł, Abdulmughni A, Brill E, Zapp J, et al. Structural basis of steroid binding and oxidation by the cytochrome P450 CYP109E1 from Bacillus megaterium. FEBS J. 2016;283:4128–48.

88. Edwards CA, Orr JC. Comparison of the 3alpha-and 20beta-hydroxysteroid dehydrogenase activities of the cortisone reductase of Streptomyces hydrogenans. Biochemistry. 1978;17:4370–6.

89. Ghosh D, Erman M, Wawrzak Z, Duax WL, Pangborn W. Mechanism of inhibition of 3 alpha, 20 beta-hydroxysteroid dehydrogenase by a licorice-derived steroidal inhibitor. Structure. 1994;2:973–80.

90. Benach J, Filling C, Oppermann UC, Roversi P, Bricogne G, Berndt KD, et al. Structure of bacterial 3beta/17beta-hydroxysteroid dehydrogenase at 1.2 A resolution: a model for multiple steroid recognition. Biochemistry. 2002;41:14659–68.
91. Knol J, Bodewits K, Hessels GI, Dijkhuizen L, van der Geize R. 3-Keto-5alpha-steroid Delta(1)-dehydrogenase from Rhodococcus erythropolis SQ1 and its orthologue in Mycobacterium tuberculosis H37Rv are highly specific enzymes that function in cholesterol catabolism. Biochem J. 2008;410:339–46.
92. Rohman A, van Oosterwijk N, Thunnissen AM, Dijkstra BW. Crystal structure and site-directed mutagenesis of 3-ketosteroid Δ1-dehydrogenase from Rhodococcus erythropolis SQ1 explain its catalytic mechanism. J Biol Chem. 2013;288:35559–68.
93. van Oosterwijk N, Knol J, Dijkhuizen L, van der Geize R, Dijkstra BW. Structure and catalytic mechanism of 3-ketosteroid-Delta4-(5α)-dehydrogenase from Rhodococcus jostii RHA1 genome. J Biol Chem. 2012;287:30975–83.
94. Li J, Vrielink A, Brick P, Blow DM. Crystal structure of cholesterol oxidase complexed with a steroid substrate: implications for flavin adenine dinucleotide dependent alcohol oxidases. Biochemistry. 1993;32:11507–15.

Molecular Determinants of Cholesterol Binding to Soluble and Transmembrane Protein Domains

Jessica Ounjian, Anna N. Bukiya, and Avia Rosenhouse-Dantsker

Abstract Cholesterol-protein interactions play a critical role in lipid metabolism and maintenance of cell integrity. To elucidate the molecular mechanisms underlying these interactions, a growing number of studies have focused on determining the crystal structures of a variety of proteins complexed with cholesterol. These include structures in which cholesterol binds to transmembrane domains, and structures in which cholesterol interacts with soluble ones. However, it remains unknown whether there are differences in the prerequisites for cholesterol binding to these two types of domains. Thus, to define the molecular determinants that characterize the binding of cholesterol to these two distinct protein domains, we employed the database of crystal structures of proteins complexed with cholesterol. Our analysis suggests that cholesterol may bind more strongly to soluble domains than to transmembrane domains. The interactions between cholesterol and the protein in both cases critically depends on hydrophobic and aromatic residues. In addition, cholesterol binding sites in both types of domains involve polar and/or charged residues. However, the percentage of appearance of the different types of polar/charged residues in cholesterol binding sites differs between soluble and transmembrane domains. No differences were observed in the conformational characteristics of the cholesterol molecules bound to soluble versus transmembrane protein domains suggesting that cholesterol is insensitive to the environment provided by the different protein domains.

Keywords Cholesterol binding · Soluble domain · Transmembrane domain · Lipid-protein interactions · Steroid binding site

J. Ounjian · A. Rosenhouse-Dantsker (✉)
Department of Chemistry, University of Illinois at Chicago, Chicago, IL, USA
e-mail: dantsker@uic.edu

A. N. Bukiya
Department of Pharmacology, The University of Tennessee HSC, Memphis, TN, USA

© Springer Nature Switzerland AG 2019
A. Rosenhouse-Dantsker, A. N. Bukiya (eds.), *Direct Mechanisms in Cholesterol Modulation of Protein Function*, Advances in Experimental Medicine and Biology 1135, https://doi.org/10.1007/978-3-030-14265-0_3

Abbreviations

CARC   Cholesterol recognition motif exhibiting an inverted CRAC orientation
       along the polypeptide chain
CCM   Cholesterol Consensus Motif
CRAC   Cholesterol Recognition Amino acid Consensus (motif)
NPC   Niemann-Pick Type C (protein)
PDB   Protein Data Bank
RCSB   Research Collaboratory for Structural Bioinformatics

1   Introduction

Cholesterol homeostasis is essential for multiple cellular functions, and is thus strictly regulated [1–5]. In recent years, a variety of proteins involved in the regulation of cholesterol trafficking and transport have been crystalized in complex with cholesterol, providing structural insights into the underlying mechanisms. These proteins are either soluble proteins or membrane proteins that possess a cholesterol binding site within a soluble domain [6–14]. These, however, are not the only crystal structures of proteins co-crystallized with cholesterol. A growing number of transmembrane proteins that are modulated by cholesterol have also been crystallized in complex with cholesterol [15–39]. These two groups identify distinct types of cholesterol binding sites. The first includes "soluble-domain cholesterol binding sites" in proteins that play a role in cholesterol homeostasis whereas the second includes "transmembrane-domain cholesterol binding sites" in proteins whose function may be affected by cholesterol.

Several different cholesterol binding motifs have been proposed to describe the prerequisites for cholesterol binding, including the cholesterol recognition amino acid consensus (CRAC) motif [40, 41], the "inverted CRAC" (CARC) motif [42], and the cholesterol consensus motif (CCM) [15]. All three cholesterol binding motifs involve hydrophobic aliphatic residues (e.g. I/L/V), aromatic residues (e.g. F/W/Y) and a positively charged residue (e.g. R/K) that can interact with the cholesterol molecule via three distinct types of interactions [43, 44]. First, leucine, valine and isoleucine can provide an adequate hydrophic environment for cholesterol. Second, the aromatic residues, which may also contribute to the hydrophobic environment, may further stabilize the interaction between the protein and the cholesterol molecule via stacking interactions. The latter type of interaction will occur when the ring structures of the cholesterol molecule and aromatic residues overlap at a van der Waals distance of separation [45]. Third, the side chains of the positively charged amino acids of the arginine and lysine residues may interact with the hydroxyl group of the cholesterol molecule via hydrogen bonding. In addition, the aromatic residues tryptophan and tyrosine can also interact with the hydroxyl group of the cholesterol molecule via their side-chains (see Fig. 1). Specifically, the tryptophan can interact with the hydroxyl group of the cholesterol through the nitrogen

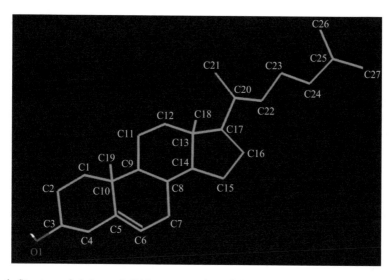

Fig. 1 Structure of cholesterol. Stick representation of the cholesterol molecule based on the cholesterol molecule bound to Beta-Cryptogein (PDB ID: 1LRI) showing the numbering of the non-hydrogen atoms in the molecule. Cholesterol is a tetracyclic molecule with a rigid fused ring system that has a polar hydrophilic hydroxyl group on one end and a flexible hydrophobic 8-carbon alkyl group on its other end

atom on the aromatic ring, and the tyrosine can interact with the hyroxyl group via the phenyl ring.

However, accumulating evidence suggests that these binding motifs do not represent all cholesterol binding pockets but rather a subset of cholesterol binding sites. First, the amphipathic helix motif identified in the squalene monooxygenase enzyme that plays a central role in choloesterol biosynthesis possesses a 12-residue segment (QFALFSDILSGL) that lacks a positively charged residue (i.e. R/K) [46]. The hydrophobic nature of the majority of the residues in this segment was found to be critical for cholesterol effect, but does not necessarily constitute a prerequisite for cholesterol binding. Second, larger cholesterol binding domains involving multiple transmembrane helices have also been implicated in cholesterol binding. An example is the sterol-sensing domain (SSD) found in the sterol regulatory element-binding protein cleavage activating protein (SCAP). The cholesterol binding site in SCAP is localized in a loop containing a 245-amino acid sequence, and binding occurs through three hydrophobic segments: (WYVGAPVAYIQQIFV), (HGCLLLSPGNFWQN), and (VSYTITLVFQ) [47]. Notably, none of these segments includes positively charged residues. Third, recent studies on cholesterol binding sites in transmembrane protein crystal structures complexed with cholesterol demonstrated that the majority of cholesterol binding sites did not include a positively charged residue [43]. Together, the above observations suggest that positively charged residues, which are an integral part of all three cholesterol binding motifs, are not a prerequisite for cholesterol binding. Fourth, analysis of crystallized cholesterol binding sites in soluble domains failed to detect any correlation between

CRAC number/distribution and the ability of non-transmembrane protein segments to bind cholesterol [48].

It remains unknown, however, whether there are differences in the molecular characteristics of "soluble-domain cholesterol binding sites" and "transmembrane-domain cholesterol binding sites". Furthermore, the dependence of the conformational characteristics of the cholesterol molecule on the type of cholesterol binding site is also unknown. With the growing number of structures showing how cholesterol binds to a variety of both transmembrane and soluble protein domains, it has become possible to define the molecular characteristics of cholesterol binding to these two different protein domains, and determine the dependence of these characteristics on the type of protein domain. Thus, utilizing the database of structures available in the Research Collaboratory for Structural Bioinformatics (RCSB) Protein Data Bank (PDB), we compared representative structures of cholesterol bound to soluble and transmembrane domains.

2 Representative Structures of Proteins Complexed with Cholesterol

The search for representative structures of proteins that include the ligand cholesterol represented by the three letter abbreviation CLR in the PDB yielded a total of 34 transmembrane structures, and 11 structures in which the cholesterol molecule was interacting with a soluble protein or a soluble domain within a transmembrane protein (see Tables 1 and 2). The resolution of the structures included in the analysis ranged from 1.45 Å to 4.3 Å. The cut-off for inclusion of protein atoms in the analysis was 4.5 Å from the cholesterol molecule heavy atoms. Within the RCSB

Table 1 Representative complexed structures with cholesterol (CLR) binding sites in soluble protein domains

PDB ID	Number of CLRs analyzed	Protein	Release date	Resolution (Å)
1LRI	1	Beta-Cryptogein	5/29/2002	1.45
1 N83	1	The retinoic acid orphan receptor (ROR) α	12/11/2002	1.63
1ZHY	1	Oxysterol binding protein Osh4	9/6/2005	1.6
3GKI	1	Niemann-Pick C 1 (NPC1)	7/14/2009	1.8
3JD8	1		6/1/2016	4.43
3N9Y	1	Cytochrome P450 11A1	6/8/2011	2.1
4BQU	1	Japanin	6/18/2014	2.36
5L7D	1	Smoothened homolog	7/20/2016	3.2
5WVR	1	Oxysterol binding protein Osh1	5/10/2017	2.2
5UPH	1	Lysosomal integral membrane protein	12/13/2017	3
6HIJ	4	ATP-binding cassette (ABC) G2	9/19/2018	3.56

Table 2 Representative complexed structures with cholesterol (CLR) binding sites in transmembrane protein domains

PDB ID	Number of CLRs analyzed	Protein	Release date	Resolution (Å)
3D4S	2	Human β2-adrenergic receptor.	6/17/2008	2.8
3PDS	1		1/12/2011	3.5
5JQH	1		7/13/2016	3.2
5D6L	1		8/17/2016	3.2
5X7D	1		8/16/2017	2.7
3KDP	1	Na$^+$, K$^+$-ATPase	2/16/2010	3.5
4HYT	2		6/26/2013	3.4
4HQJ	2		10/2/2013	4.3
4RET	2		1/28/2015	4.0
5AVQ	1		9/2/2015	2.6
4XE5	2		3/9/2016	3.9
3 AM6	2	Rhodopsin AR2	7/6/2011	3.2
4NC3	1	5-HT$_{2B}$ serotonin receptor	12/18/2013	2.8
4OR2	3	Metabotropic glutamate receptor 1	3/19/2014	2.8
4NTJ	2	P2Y$_{12}$ purinergic receptor	3/26/2014	2.62
4PXZ	1		4/30/2014	2.5
4XT1	2	GPCR homolog US28	3/4/2015	2.89
4XNV	1	P2Y$_1$ purinergic receptor	4/1/2015	2.2
4XNX	2	Dopamine transporter	5/13/2015	3.0
5C1M	1	μ-opioid receptor	8/5/2015	2.1
5I6X	1	ts3 serotonin transporter	4/13/2016	3.14
5SY1	1	STRA6 receptor	8/24/2016	3.9
5TCX	1	Tetraspanin CD81	11/9/2016	2.95
5LWE	1	CC chemokine receptor type 9 (CCR9)	12/7/2016	2.8
5WB2	2		6/13/2018	3.5
5UVI	2	A2a adenosine receptor	5/24/2017	3.2
5N2R	1		7/26/2017	2.8
5MZJ	1		7/26/2017	2.0
5XRA	1	Cannabinoid receptor 1 (CB$_1$)	7/12/2017	2.8
5X93	1	Endothelin receptor B (ET$_B$)	8/16/2017	2.2
6BHU	3	Multidrug resistance protein 1 (MRP1)	12/27/2017	3.14
6B73	1	κ-opioid receptor	1/17/2018	3.1
5OQT	1	Cationic amino acid transporter (CAT)	2/14/2018	2.86
6CO7	1	nvTRPM2 channel	5/16/2018	3.07

database of structures, there were some sets of structures in which the protein was crystallized in different conditions (e.g. different ligands, different agonists, etc.). Also, some proteins were crystallized with cholesterol in different occasions (e.g. at different times, by different laboratories, etc.). If the residues that formed the cholesterol binding site were very similar (>90% identity), only one of the structures was included in the analysis. Since several structures included more than one cholesterol molecule, all unique (non-duplicate) binding regions were included in the analysis resulting in a total of 49 cholesterol-transmembrane protein domain binding sites and 14 cholesterol-soluble protein domain binding sites. Together, these sets of structures are representative of the currently available cholesterol-bound proteins within the PDB.

3   Differential Molecular Characteristics of Cholesterol Binding Sites in Transmembrane Versus Soluble Protein Domains

Cholesterol binding sites in transmembrane domains usually face the lipids of the plasma membrane. In contrast, in soluble domains, the cholesterol molecule is engulfed by protein residues that shield it from the surrounding hydrophilic environment (Fig. 2a). Accordingly, whereas transmembrane-domain binding sites are mostly "pocket-shaped", soluble-domain binding sites are primarily "tunnel-shaped". Thus, cholesterol binding sites in soluble domains involve a significantly higher number of protein residues compared to cholesterol binding sites in transmembrane domains ($P = 4.8 \times 10^{-11}$; two sample independent t-Test). The average total number of residues located within 4.5 Å from the cholesterol molecule in the soluble domains analyzed was 17.6 ± 1.3, whereas in the transmembrane domains, the overall number of residues within the same distance was 10.0 ± 0.3.

To further define the protein environment of the cholesterol molecule in transmembrane and soluble domains, we determined the average number of residues that interacted with each of the cholesterol atoms in each of the binding sites in the structures listed in Table 1 (soluble) and Table 2 (transmembrane). These numbers are depicted in Fig. 2b. In alignment with the increase in the total number of residues that surround the cholesterol molecule, the number of protein residues located within 4.5 Å from each of the cholesterol atoms in soluble domains is significantly higher compared to transmembrane domains ($P = 1.4 \times 10^{-6}$; two sample independent t-Test). The average number of protein residues within 4.5 Å from a non-hydrogen cholesterol atom is 2.0 ± 0.1 in cholesterol binding sites in soluble domains, and 1.1 ± 0.1 in cholesterol binding sites in transmembrane domains. Interestingly, the location of the maximal average proximal number of residues in transmembrane cholesterol binding sites differs from its location in soluble domains cholesterol binding sites. In cholesterol binding sites in transmembrane domains, the largest average number of proximal residues is observed at the hydroxyl group and the adjacent C4 atom of the cholesterol molecule. In contrast, the largest average number of proximal residues in cholesterol binding sites in soluble domains is observed at the other end of the cholesterol molecule, at the tip of its tail at C26/C27. This difference suggests that protein residues are used to shield different regions of the cholesterol molecule depending on the environment. Specifically, protein residues in transmembrane domains shield the hydroxyl group from the surrounding hydrophobic environment of the membrane, whereas in solu-

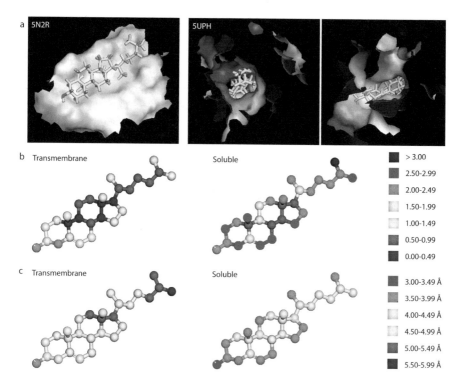

Fig. 2 Differential characteristics of cholesterol binding sites in soluble and transmembrane domains. (**a**) Electrostatic surface representation of examples of cholesterol bound to transmembrane (left, A2a adenosine receptor, PDB ID: 5N2R) and soluble (right two panels, lysosomal integral membrane protein, PDB ID: 5UPH) protein domains. In comparison to the often seen "pocket-shaped" transmembrane-domain binding site (left panel), the soluble-domain binding site is commonly "tunnel-shaped" (the middle panel shows the cholesterol molecule from the hydroxyl end, and the right panel shows the cholesterol molecule from the tail). The surface is colored according to the electrostatic potential as calculated in PyMol. Red indicates positively charged, blue indicates negatively charged and white indicates neutral residues. (**b**) Average number of residues interacting with each of the cholesterol atoms in each of the binding sites in the soluble and transmembrane domains listed in Table 1 (soluble) and Table 2 (transmembrane) within 4.5 Å. (**c**) Average minimal distances between each of the cholesterol atoms and the protein residues in cholesterol binding sites in soluble and transmembrane protein domains listed in Table 1 (soluble) and Table 2 (transmembrane). To take into account all the cholesterol atoms, no distance cut-off was employed

ble domains, the hydrophobic tail of the cholesterol molecule is shielded from the hydrophilic environment of the surrounding solvent.

With a lower number of protein residues interacting with the cholesterol molecule in transmembrane domains compared to soluble domains, the interaction between cholesterol and transmembrane domains is likely to be weaker than with soluble domains. Further support of this notion is provided through a comparison of the average minimal distances between each of the atoms in the cholesterol molecule and their surrounding protein residues in the protein database of structures analyzed (Tables 1 and 2). Not all cholesterol atoms are within 4.5 Å from a protein residue. Thus, to account for the minimal distances between each of the atoms in the cholesterol molecule and their most proximal surrounding protein residue, no cut-off was used in this analysis. As evident in Fig. 2c, the average minimal distances between each of the cholesterol atoms and the protein residues are significantly shorter in soluble binding sites than in transmembrane binding sites. Consistent with this observation, the average minimal distance between the cholesterol molecule and the closest protein residue is significantly shorter in soluble domains than in transmembrane domains ($P = 1.77 \times 10^{-6}$; two sample independent t-Test). Specifically, the overall average minimal distance is 4.07 ± 0.11 Å in cholesterol binding sites in soluble domains and 4.76 ± 0.06 Å in cholesterol binding sites in transmembrane domains.

Next, to compare the types of residues that form cholesterol binding sites in soluble protein domains (Table 1) with those in transmembrane protein domains (Table 2), we plotted the percentage of appearance of each of the 20 different amino acid residues within 4.5 Å from each non-hydrogen atom in the cholesterol molecule for both domain types. The results are depicted in Fig. 3a, b. For each cholesterol atom, the amino acids are shown as bars, organized from left to right according to the order and color in the key. The height of each bar indicates the percentage of appearance of the amino acid residue of the corresponding type within 4.5 Å from the cholesterol atom. To facilitate the assessment of the differences between the molecular composition of cholesterol binding sites in soluble and transmembrane domains, we calculated the differences between the percentage of appearance of each of the amino acid residues in cholesterol binding sites in soluble and transmembrane domains. These differences are depicted in Fig. 3c. In this figure, positive values indicate higher residue appearance in cholesterol binding sites in soluble domains compared to transmembrane domains whereas negative values denote higher residue appearance in cholesterol binding sites in transmembrane domains compared to soluble domains. The overall percentage of appearance of each of the 20 amino acid residues within cholesterol binding sites in soluble and transmembrane domains is shown in Fig. 4.

Examination of the information in Figs. 3 and 4 leads to several key observations. First, all cholesterol binding sites in both soluble and transmembrane domains involve multiple isoleucines, leucines and valines. These hydrophobic aliphatic residues appear within 4.5 Å from the vast majority of cholesterol atoms. The only exception is C13 of the cholesterol molecule, which does not interact with this type of residues in all the soluble-domain cholesterol binding sites included in the analysis (Table 1). Furthermore, while leucines were observed in the vicinity of C13 in transmembrane-domain cholesterol binding sites, their appearance did not exceed 10% of the residues located within 4.5 Å from C13. Instead, this cholesterol atom, which is located within the ring structure of the cholesterol molecule, interacted primarily with aromatic residues in cholesterol binding sites in both soluble and transmembrane domains. The consistent representation of this group of hydrophobic aliphatic residues of isoleucines, leucines and valines in cholesterol binding sites indicates that this type of residues is critical for forming a favorable hydrophobic environment for cholesterol binding in both types of domains. This hydrophobic environment is further enhanced by alanine, glycine, methionine and proline in both soluble and transmembrane cholesterol binding sites. Together, as depicted in

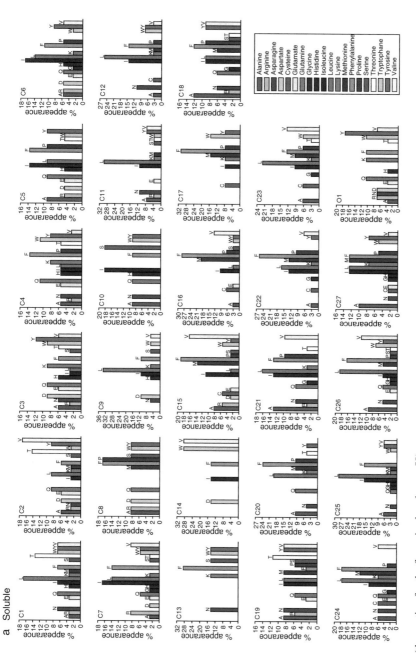

Fig. 3 (the caption for the figure is placed on p. 58)

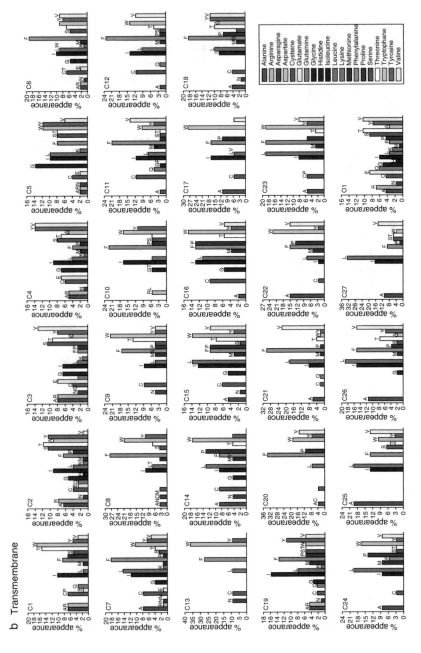

Fig. 3 (the caption for the figure is placed on p. 58)

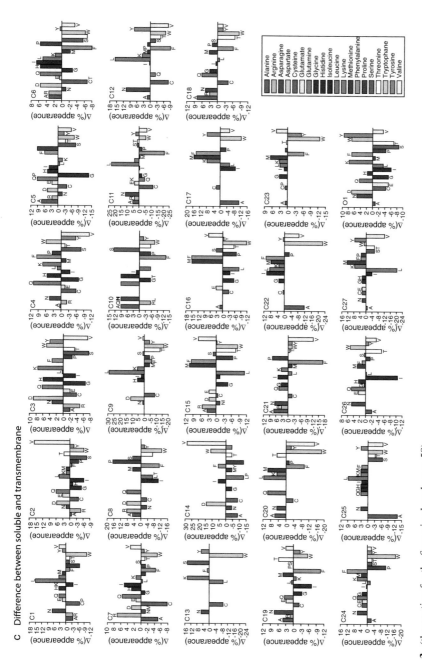

Fig. 3 (the caption for the figure is placed on p. 58)

Fig. 3 Percentage of appearance of amino acid residues in proximity to the non-hydrogen cholesterol atoms in cholesterol binding sites in soluble and transmembrane domains. Percentage of appearance of each of the 20 different amino acid residues within 4.5 Å from each non-hydrogen atom in the cholesterol molecule in the cholesterol binding sites in the (**a**) soluble and (**b**) transmembrane domains listed in Table 1 (soluble) and Table 2 (transmembrane). The amino acid residues are shown as bars, organized from left to right according to the order and color in the key. A separate bar graph is plotted for each cholesterol non-hydrogen atom. The height of each bar indicates the percentage of appearance of the amino acid residue of the corresponding type within 4.5 Å from the cholesterol atom. (**c**) Differences between the percentage of appearance of each of the amino acid residues in cholesterol binding sites in soluble and transmembrane domains. Positive values indicate higher residue appearance in cholesterol binding sites in soluble domains compared to transmembrane domains whereas negative values denote higher residue appearance in cholesterol binding sites in transmembrane domains compared to soluble domains

Fig. 4a, b, these seven hydrophobic residues comprise just over 50% of the residues that form cholesterol binding sites in both soluble and transmembrane domains.

A second key observation is that both soluble-domain and transmembrane-domain cholesterol binding sites involve the hydrophobic aromatic residues phenylalanine, tryptophan, and tyrosine. As evident in Fig. 3a, b, these residues often appear within 4.5 Å from each of the cholesterol atoms. This observation highlights the importance of this type of residues for cholesterol binding in both soluble and transmembrane domains. Together, they comprise approximately 23% of the residues that form cholesterol binding sites in both types of domains (Fig. 4a, b). These aromatic residues can stabilize the interactions between the protein and cholesterol in several complementary ways. First, they can form stacking interactions with the rigid ring structure of the cholesterol molecule. Second, these hydrophobic residues can also contribute to the hydrophobic environment required for cholesterol binding. Third, the side-chains of two of these amino acid residues, tyrosine and tryptophan, include a hydrophilic group (an ionizable hydroxyl group on the phenyl ring of the tyrosine and a nitrogen in the aromatic indole ring of the tryptophan) that can interact with the hydroxyl group of cholesterol via their side-chains.

The third key observation evident in Figs. 3 and 4 is the relatively limited appearance of the positively charged residues arginine and lysine within the database of structures analyzed (Tables 1 and 2). In the vicinity of the majority of the atoms of the cholesterol molecule, the combined percentage of appearance of arginines and lysines is generally higher in soluble domains than in transmembrane domains. Consistent with this observation, together with the positively charged residue histidine, the overall percentage of appearance of these residues is reduced by almost 60% in cholesterol binding sites in transmembrane domains compared to soluble domains (Fig. 4a, b). The reduced appearance of these positively charged residues in cholesterol binding sites located in transmembrane domains may be attributed to the overall reduced probability of positively charged residues in transmembrane protein segments embedded in the hydrophobic environment of the plasma membrane. Notably, all three positively charged residues are absent from the proximal vicinity of the tail atoms of the cholesterol molecule in cholesterol binding sites in

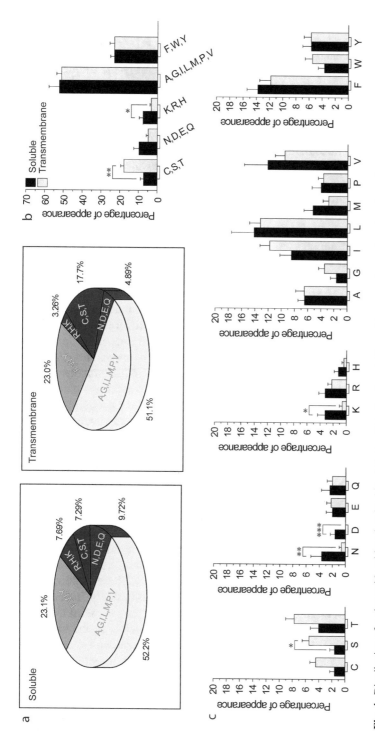

Fig. 4 Distribution of amino acid residues involved in cholesterol binding in soluble and transmembrane domains. Overall percentage of appearance of each of the 20 amino acid residues within 4.5 Å from the cholesterol molecule in cholesterol binding sites in soluble and transmembrane domains listed in Table 1 (soluble) and Table 2 (transmembrane). The combined percentage of appearance of different types of amino acid residues is depicted in (**a**) pie charts and (**b**) a bar graph. The overall percentage of appearance of each of the 20 amino acid residues is depicted in (**c**) bar graphs, and organized in groups based on the type of residue

transmembrane domains, which is not the case in cholesterol binding sites in soluble proteins.

On the other hand, in the vicinity of the hydroxyl group of the cholesterol molecule, the positively charged residues arginine and/or lysine and/or histidine are found in some cholesterol binding sites in both soluble and transmembrane domains (Fig. 3). However, whereas the percentage of appearance of arginine and histidine is comparable in the vicinity of the oxygen atom of the hydroxyl group of the cholesterol molecule in both types of domains, the percentage of appearance of lysine is higher in cholesterol binding sites in soluble domains. As we have shown in an earlier study, even in cases in which there is an arginine or a lysine within 5 Å from the cholesterol molecule, it does not necessarily interact with the hydroxyl group of the cholesterol molecule [43]. In some cases, the interaction of the hydroxyl group with a positively charged amino acid is replaced by an interaction with surrounding water molecules or with the hydroxyl group of an adjacent cholesterol molecule suggesting that positively charged residues are not a prerequisite for cholesterol binding [43]. As Fig. 3 indicates, in the majority of cholesterol binding sites, these positively charged residues are replaced by other types residues that interact with the cholesterol hydroxyl group either via their side-chain or backbone functional groups. While almost any residue seems to be able to replace these positively charged residues, one group stands out. This group includes the polar cysteine, serine, and threonine residues, which exhibit an increased percentage of appearance in cholesterol binding sites in transmembrane domains compared to soluble domains (Figs. 3 and 4). This observation may be attributed to the hydrogen bonding that can be formed between the side-chains of these residues and the hydroxyl group of the cholesterol molecule. Consistent with their increased appearance in proximity to the hydroxyl group in transmembrane-domain cholesterol binding sites, the percentage of appearance of this group of residues within 4.5 Å from proximal carbon atoms to the hydroxyl group of the cholesterol molecule (C1-C6) is also significantly higher in cholesterol binding sites in transmembrane domains compared to soluble domains (Fig. 3).

In addition, the percentage of appearance of asparagine and aspartate in cholesterol binding sites is significantly different in soluble versus transmembrane domains. Both residues are significantly reduced or absent from the cholesterol binding sites in transmembrane domains. Thus, as summarized in Fig. 4, the decrease in the percentage of appearance of these residues along with the decrease in the positively charged residues arginine, lysine and histidine, in cholesterol binding sites formed in transmembrane domains compared to soluble domains, is countered by an increase in cysteine, serine and threonine. In contrast, there is no significant change in the percentage of appearance of hydrophobic and aromatic residues in cholesterol binding sites in the two types of domains, transmembrane and soluble. This suggests that while cholesterol binding sites can accommodate different types of polar and charged residues, they require a rather constant percentage of hydrophobic and aromatic residues.

4   The Structural Characteristics of the Cholesterol Molecule Are Comparable in Soluble and Transmembrane Protein Domains

A remaining question is whether the conformation of the cholesterol molecule itself depends on the type of domain, i.e. soluble versus transmembrane, to which it binds. Whereas the ring system of the cholesterol molecule is relatively rigid, its tail possesses intrinsic flexibility, which may have functional implications. For example, rotation of the C17-C20-C22-C23 dihedral angle of cholesterol was suggested to occur during the transfer of a cholesterol molecule between the two lysosomal Niemann-Pick C (NPC) proteins that are required for cholesterol export from the lysosomes, i.e. from NPC2 to NPC1. Specifically, simulations suggested that this dihedral angle is 71.6° when cholesterol is in the NPC2 pocket but −157.3° (202.7°) when it is bound to the NPC1 N-terminal domain [49, 50]. The latter is very close to the value of −163.9° (196.1°) observed in the crystal structures of NPC1 complexed with cholesterol (PDB IDs 3GKI and 3JD8) or to the value of −162.2° (197.8°) observed for NPC1 structures complexed with 25-hydroxycholesterol (PDB ID 3GKJ). In contrast, the former differs from the values of 174.3° (PDB ID 5KWY) and −164.5° (195.5°) (PDB ID 2HKA) observed for NPC2 complexed with the cholesterol derivative cholesterol sulfate. It was thus suggested that rotation of C17-C20-C22-C23 occurs during an intermediate transition state [50]. Using a quantum mechanical description of the cholesterol molecule and a molecular mechanics force field to describe NPC1 and NPC2, the energy barrier for rotation of this dihedral angle was estimated to be approximately 22 kcal/mol [50].

We thus compared the distribution of the C17-C20-C22-C23 dihedral angle in cholesterol molecules bound to the soluble and transmembrane domains listed in Tables 1 and 2. As Figs. 5a, b show, the C17-C20-C22-C23 dihedral angle acquires two different values independent of whether cholesterol binds to a soluble or transmembrane domain. Together, these distinct distributions are centered around (I) 77° and around (II) 193°. Among the cholesterol molecules bound to the soluble domains analyzed, the averages are (I) 64 ± 9° and (II) 189 ± 6°, and among the cholesterol molecules bound to the transmembrane domains, the averages are (I) 83 ± 6° and (II) 194 ± 5°. There is no significant difference between the distributions of this cholesterol dihedral angle in soluble- and transmembrane-domain binding sites ($P = 0.1$ for the distributions around (I), and $P = 0.6$ for the distributions around (II); two sample independent t-Test).

To further compare the conformational profiles of cholesterol molecules bound to soluble versus transmembrane protein domains, we downloaded structural data from the database of structures listed in Tables 1 and 2 into Molecular Operating Environment (MOE, Chemical Computing Group), and the radius of gyration, critical packing parameter, and potential energy were calculated for each of the two cholesterol databases using a built-in function in MOE. Whereas the radius of gyration describes the distribution of atoms around the overall molecular axis [51], the critical packing parameter is derived based on the relative area occupied by

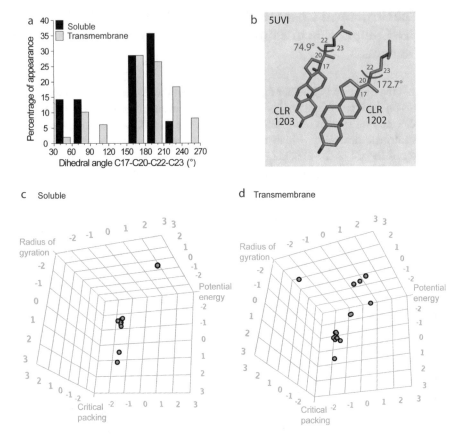

Fig. 5 The structural characteristics of cholesterol is comparable in soluble and transmembrane protein domains. (**a**) Histogram of the percentage of appearance of the possible values of the C17-C20-C22-C23 dihedral angle in the soluble and transmembrane domains listed in Table 1 (soluble) and Table 2 (transmembrane). (**b**) Examples of representative tail conformations in the two clusters of the possible values of the C17-C20-C22-C23 dihedral angle in soluble and transmembrane domain. The depicted cholesterol structures are based on two of the cholesterol molecules bound to A2a adenosine receptor in PDB ID 5UVI. (**c, d**) Conformational profile of cholesterol molecules bound to soluble versus transmembrane protein domains depicting the potential energy, radius of gyration and critical packing. The parameters were calculated by using a built in function in Molecular Operating Environment (MOE, Chemical Computing Group). Prior to inclusion into the calculation set, structural data were visually inspected for the presence of the cholesterol-characteristic double bond. PDB entries that failed to produce the presence of a double bond were excluded from further analysis. When several satisfactory entries existed for a given protein site, the structure with the highest resolution was used. Calculations were performed using the following PDB ID entries. For soluble protein domains, the PDB ID entries used were: 1LRI, 1N83, 1ZHY, 3GKI, 3N9Y, 4BOE, 5L7D, 5UOH, 5WVR, 6HIJ. For transmembrane domains, the following PDB entries were used: 2RH1, 2ZHE, 3AM6, 4DKL, 4NC3, 4OR2, 4PXZ, 4XNV, 4XNX, 4XT1, 5I6X, 5IU4, 5LWE, 5OQT, 5TCX, 5XRA, 6BHU, 6CO7

hydrophobic versus hydrophilic chemical moieties in amphiphiles [52]. Last, the potential energy provides an estimate on how distant the system is from the energetic minimum [53]. Despite a somewhat higher variability in these parameters within the database of cholesterol molecules bound to transmembrane protein segments (Fig. 5c vs. d), there was no significant difference between the two databases ($P = 0.3880$, 0.1553, and 0.0717 for the radius of gyration, the critical packing parameter, and the potential energy, respectively; two-tail Mann-Whitney tests). This suggests that the nature of the binding partner, either soluble or transmembrane protein domain, does not play a critical role in these three shaping descriptors of cholesterol as a ligand.

5   Concluding Remarks

In this chapter we explored the molecular characteristics of cholesterol binding sites in soluble and transmembrane protein domains, and the effect of soluble versus membrane environments on the conformation of the flexible tail of the cholesterol molecule as well as on several indicators of conformational alterations (radius of gyration, critical packing, and potential energy).

Our analysis suggests a comparable critical role of hydrophobic aliphatic and aromatic residues in the interaction of cholesterol with both types of protein domains. The primary differences among the types of protein residues that interact with cholesterol in soluble versus transmembrane domains lie in the types of polar and charged residues involved in interactions with the cholesterol molecule, especially with its hydroxyl group. A significant percentage of the positively charged residues arginine, lysine and histidine, as well as of asparagine and aspartate that appear in cholesterol binding sites in soluble domains, is replaced by the polar cysteines, serines and threonines in cholesterol binding sites in transmembrane domains. Additional significant differences lie in the average number of protein residues that interact with the cholesterol molecule, and the average minimal distance between these residues and the cholesterol molecule. Compared to binding sites in soluble domains, cholesterol binding sites in transmembrane proteins involve a smaller number of protein residues, which, on the average, are more distant from the cholesterol molecule. These differences suggest that the interaction between the cholesterol molecule and protein residues is stronger in soluble domains than in transmembrane domains. Yet, despite these differences, the conformation of the cholesterol molecule is unaffected by the type of domain to which it binds suggesting that cholesterol is insensitive to the differences, and that the comparable percentage of hydrophobic aliphatic and aromatic residues in the binding sites in both types of domains plays a dominant role in facilitating protein-cholesterol interactions.

References

1. Berg JM, Tymczko JL, Stryer L. The complex regulation of cholesterol biosynthesis takes place at several levels. In: Biochemistry. 7th ed. New York. Section 26.3: W.H. Freeman; 2012. p. 770–9.
2. Afonso SM, Machado RM, Lavrador MS, Quintao ECR, Moore KJ, Lottenberg AM. Molecular pathways underlying cholesterol homeostasis. Nutrients. 2018;10:E760.
3. Maxfield FR, van Meer G. Cholesterol, the central lipid of mammalian cells. Curr Opin Cell Biol. 2010;22:422–9.
4. Zhang J, Liu Q. Cholesterol metabolism and homeostasis in the brain. Protein Cell. 2015;6:254–64.
5. Jeske DJ, Dietschy JM. Regulation of rates of cholesterol synthesis in vivo in the liver and carcass of the rat measured using [3H] water. J Lipid Res. 1980;21:364–76.
6. Lascombe MB, Ponchet M, Venard P, Milat ML, Blein JP, Prangé T. The 1.45A resolution structure of the cryptogein-cholesterol complex: a close-up view of a sterol carrier protein (SCP) active site. Acta Crystallogr D Biol Crystallogr. 2002;58:1442–7.
7. Kallen JA, Schlaeppi JM, Bitsch F, Geisse S, Geiser M, Delhon I, Fournier B. X-ray structure of hRORα LBD at 1.63A: structural and functional data that cholesterol or a cholesterol derivative is the natural ligand of RORα. Structure. 2002;10:1697–707.
8. Gong X, Qian H, Zhou X, Wu J, Wan T, Cao P, Huang W, Zhao X, Wang X, Wang P, Shi Y, Gao GF, Zhou Q, Yan N. Structural insights into the Niemann-Pick C1 (NPC1)-mediated cholesterol transfer and Ebola infection. Cell. 2016;165:1467–78.
9. Strushkevich N, MacKenzie F, Cherkesova T, Grabovec I, Usanov S, Park HW. Structural basis for pregnenolone biosynthesis by the mitochondrial monooxygenase system. Proc Natl Acad Sci U S A. 2011;108:10139–43.
10. Roversi P, Johnson S, Preston SG, Nunn MA, Paesen GC, Austyn JM, Nuttall PA, Lea SM. Structural basis of cholesterol binding by a novel clade of dendritic cell modulators from ticks. Sci Rep. 2017;7:16057.
11. Byrne EFX, Sircar R, Miller PS, Hedger G, Luchetti G, Nachtergaele S, Tully MD, Mydock-McGrane L, Covey DF, Rambo RP, Sansom MSP, Newstead S, Rohatgi R, Siebold C. Structural basis of smoothened regulation by its extracellular domains. Nature. 2016;535(7613):517–22.
12. Manik MK, Yang H, Tong J, Im YJ. Structure of yeast OSBP-related protein osh1 reveals key determinants for lipid transport and protein targeting at the nucleus-vacuole junction. Structure. 2017;25:617–629.e3.
13. Conrad KS, Cheng TW, Ysselstein D, Heybrock S, Hoth LR, Chrunyk BA, Am Ende CW, Krainc D, Schwake M, Saftig P, Liu S, Qiu X, Ehlers MD. Lysosomal integral membrane protein-2 as a phospholipid receptor revealed by biophysical and cellular studies. Nat Commun. 2017;8:1908.
14. Jackson SM, Manolaridis I, Kowal J, Zechner M, Taylor NMI, Bause M, Bauer S, Bartholomaeus R, Bernhardt G, Koenig B, Buschauer A, Stahlberg H, Altmann KH, Locher KP. Structural basis of small-molecule inhibition of human multidrug transporter ABCG2. Nat Struct Mol Biol. 2018;25:333–40.
15. Hanson MA, Cherezov V, Griffith MT, Roth CB, Jaakola VP, Chien EYT, Velasquez J, Kuhn P, Stevens RC. A specific cholesterol binding site is established by the 2.8Å structure of the human β-adrenergic receptor. Structure. 2008;16:897–905.
16. Cherezov V, Rosenbaum DM, Hanson MA, Rasmussen SG, Thian FS, Kobilka TS, Choi HJ, Kuhn P, Weis WI, Kobilka BK, Stevens RC. High-resolution crystal structure of an engineered human beta2-adrenergic G protein-coupled receptor. Science. 2007;318:1258–65.
17. Morth JP, Pedersen BP, Toustrup-Jensen MS, Sørensen TL, Petersen J, Andersen JP, Vilsen B, Nissen P. Crystal structure of the sodium-potassium pump. Nature. 2007;450:1043–9.
18. Shinoda T, Ogawa H, Cornelius F, Toyoshima C. Crystal structure of the sodium-potassium pump at 2.4 A resolution. Nature. 2009;459:446–50.

19. Wada T, Shimono K, Kikukawa T, Hato M, Shinya N, Kim SY, Kimura-Someya T, Shirouzu M, Tamogami J, Miyauchi S, Jung KH, Kamo N, Yokoyama S. Crystal structure of the eukaryotic light-driven proton-pumping rhodopsin, Acetabularia rhodopsin II, from marine alga. J Mol Biol. 2011;411:986–98.

20. Liu W, Wacker D, Gati C, Han GW, James D, Wang D, Nelson G, Weierstall U, Katritch V, Barty A, Zatsepin NA, Li D, Messerschmidt M, Boutet S, Williams GJ, Koglin JE, Seibert MM, Wang C, Shah ST, Basu S, Fromme R, Kupitz C, Rendek KN, Grotjohann I, Fromme P, Kirian RA, Beyerlein KR, White TA, Chapman HN, Caffrey M, Spence JC, Stevens RC, Cherezov V. Serial femtosecond crystallography of G protein-coupled receptors. Science. 2013;342:1521–4.

21. Wu H, Wang C, Gregory KJ, Han GW, Cho HP, Xia Y, Niswender CM, Katritch V, Meiler J, Cherezov V, Conn PJ, Stevens RC. Structure of a class C GPCR metabotropic glutamate receptor 1 bound to an allosteric modulator. Science. 2014;344:58–64.

22. Zhang K, Zhang J, Gao ZG, Zhang D, Zhu L, Han GW, Moss SM, Paoletta S, Kiselev E, Lu W, Fenalti G, Zhang W, Müller CE, Yang H, Jiang H, Cherezov V, Katritch V, Jacobson KA, Stevens RC, Wu B, Zhao Q. Structure of the human P2Y12 receptor in complex with an antithrombotic drug. Nature. 2014;509:115–8.

23. Burg JS, Ingram JR, Venkatakrishnan AJ, Jude KM, Dukkipati A, Feinberg EN, Angelini A, Waghray D, Dror RO, Ploegh HL, Garcia KC. Structural biology. Structural basis for chemokine recognition and activation of a viral G protein-coupled receptor. Science. 2015;347:1113–7.

24. Zhang D, Gao ZG, Zhang K, Kiselev E, Crane S, Wang J, Paoletta S, Yi C, Ma L, Zhang W, Han GW, Liu H, Cherezov V, Katritch V, Jiang H, Stevens RC, Jacobson KA, Zhao Q, Wu B. Two disparate ligand-binding sites in the human P2Y1 receptor. Nature. 2015;520:317–21.

25. Penmatsa A, Wang KH, Gouaux E. X-ray structure of dopamine transporter elucidates antidepressant mechanism. Nature. 2013;503:85–90.

26. Penmatsa A, Wang KH, Gouaux E. X-ray structures of Drosophila dopamine transporter in complex with nisoxetine and reboxetine. Nat Struct Mol Biol. 2015;22:506–8.

27. Huang W, Manglik A, Venkatakrishnan AJ, Laeremans T, Feinberg EN, Sanborn AL, Kato HE, Livingston KE, Thorsen TS, Kling RC, Granier S, Gmeiner P, Husbands SM, Traynor JR, Weis WI, Steyaert J, Dror RO, Kobilka BK. Structural insights into μ-opioid receptor activation. Nature. 2015;524:315–21.

28. Coleman JA, Green EM, Gouaux E. X-ray structures and mechanism of the human serotonin transporter. Nature. 2016;532:334–9.

29. Chen Y, Clarke OB, Kim J, Stowe S, Kim YK, Assur Z, Cavalier M, Godoy-Ruiz R, von Alpen DC, Manzini C, Blaner WS, Frank J, Quadro L, Weber DJ, Shapiro L, Hendrickson WA, Mancia F. Structure of the STRA6 receptor for retinol uptake. Science. 2016;353:aad8266.

30. Zimmerman B, Kelly B, McMillan BJ, Seegar TCM, Dror RO, Kruse AC, Blacklow SC. Crystal structure of a full-length human tetraspanin reveals a cholesterol-binding pocket. Cell. 2016;167:1041–1051.e11.

31. Oswald C, Rappas M, Kean J, Doré AS, Errey JC, Bennett K, Deflorian F, Christopher JA, Jazayeri A, Mason JS, Congreve M, Cooke RM, Marshall FH. Intracellular allosteric antagonism of the CCR9 receptor. Nature. 2016;540:462–5.

32. Martin-Garcia JM, Conrad CE, Nelson G, Stander N, Zatsepin NA, Zook J, Zhu L, Geiger J, Chun E, Kissick D, Hilgart MC, Ogata C, Ishchenko A, Nagaratnam N, Roy-Chowdhury S, Coe J, Subramanian G, Schaffer A, James D, Ketwala G, Venugopalan N, Xu S, Corcoran S, Ferguson D, Weierstall U, Spence JCH, Cherezov V, Fromme P, Fischetti RF, Liu W. Serial millisecond crystallography of membrane and soluble protein microcrystals using synchrotron radiation. IUCrJ. 2017;4:439–54.

33. Cheng RKY, Segala E, Robertson N, Deflorian F, Doré AS, Errey JC, Fiez-Vandal C, Marshall FH, Cooke RM. Structures of human A$_1$ and A$_{2A}$ adenosine receptors with xanthines reveal determinants of selectivity. Structure. 2017;25:1275–85.

34. Hua T, Vemuri K, Nikas SP, Laprairie RB, Wu Y, Qu L, Pu M, Korde A, Jiang S, Ho JH, Han GW, Ding K, Li X, Liu H, Hanson MA, Zhao S, Bohn LM, Makriyannis A, Stevens

RC, Liu ZJ. Crystal structures of agonist-bound human cannabinoid receptor CB1. Nature. 2017;547:468–71.

35. Shihoya W, Nishizawa T, Yamashita K, Inoue A, Hirata K, Kadji FMN, Okuta A, Tani K, Aoki J, Fujiyoshi Y, Doi T, Nureki O. X-ray structures of endothelin ETB receptor bound to clinical antagonist bosentan and its analog. Nat Struct Mol Biol. 2017;24:758–64.

36. Johnson ZL, Chen J. ATP binding enables substrate release from multidrug resistance protein 1. Cell. 2018;172:81–89.e10.

37. Che T, Majumdar S, Zaidi SA, Ondachi P, McCorvy JD, Wang S, Mosier PD, Uprety R, Vardy E, Krumm BE, Han GW, Lee MY, Pardon E, Steyaert J, Huang XP, Strachan RT, Tribo AR, Pasternak GW, Carroll FI, Stevens RC, Cherezov V, Katritch V, Wacker D, Roth BL. Structure of the nanobody-stabilized active state of the kappa opioid receptor. Cell. 2018;172:55–67.

38. Jungnickel KEJ, Parker JL, Newstead S. Structural basis for amino acid transport by the CAT family of SLC7 transporters. Nat Commun. 2018;9:550.

39. Zhang Z, Tóth B, Szollosi A, Chen J, Csanády L. Structure of a TRPM2 channel in complex with Ca^{2+} explains unique gating regulation. elife. 2018;7:e36409.

40. Epand RM. Cholesterol and the interaction of proteins with membrane domains. Prog Lipid Res. 2006;45:279–94.

41. Li H, Papadopoulos V. Peripheral-type benzodiazepine receptor function in cholesterol transport. Identification of a putative cholesterol recognition/interaction amino acid sequence and consensus pattern. Endocrinology. 1998;139:4991–7.

42. Fantini J, Barrantes FJ. How cholesterol interacts with membrane proteins: an exploration of cholesterol-binding sites including CRAC, CARC and tilted domains. Front Physiol. 2013;4:31.

43. Rosenhouse-Dantsker A. Insights into the molecular requirements for cholesterol binding to ion channels. Curr Top Membr. 2017;80:187–208.

44. Singh AK, McMillan J, Bukiya AN, Burton B, Parrill AL, Dopico AM. Multiple cholesterol recognition/interaction amino acid consensus (CRAC) motifs in cytosolic C tail of Slo1 subunit determine cholesterol sensitivity of Ca^{2+}- and voltage-gated K^+_- (BK) channels. J Biol Chem. 2012;287:20509–21.

45. Maresca M, Derghal A, Caravagna C, Dudin S, Fantini J. Controlled aggregation of adenine by sugars: physicochemical studies, molecular modelling simulations of sugar-aromatic CH-pi stacking interactions, and biological significance. Phys Chem Chem Phys. 2008;10:2792–800.

46. Chua NK, Howe V, Jatana N, Thukral L, Brown AJ. A conserved degron containing an amphipathic helix regulates the cholesterol-mediated turnover of human squalene monooxygenase, a rate-limiting enzyme in cholesterol synthesis. J Biol Chem. 2017;292:19959–73.

47. Motamed M, Zhang Y, Wang ML, Seemann J, Kwon HJ, Goldstein JL, Brown MS. Identification of luminal loop 1 of Scap protein as the sterol sensor that maintains cholesterol homeostasis. J Biol Chem. 2011;286(20):18002–12.

48. Bukiya AN, Dopico AM. Common structural features of cholesterol binding sites in crystallized soluble proteins. J Lipid Res. 2017;58:1044–54.

49. Estiu G, Khatri N, Wiest O. Computational studies of the cholesterol transport between NPC2 and the N-terminal domain of NPC1 (NPC1(NTD)). Biochemistry. 2013;52:6879–91.

50. Elghobashi-Meinhardt N. Niemann–pick type C disease: a QM/MM study of conformational changes in cholesterol in the NPC1(NTD) and NPC2 binding pockets. Biochemistry. 2014;53:6603–14.

51. Lei J. Probability distribution of the radius of gyration of freely jointed chains. J Chem Phys. 2010;133:104903.

52. Khalil RA, Zarari AA. Theoretical estimation of the critical packing parameter of amphiphilic self-assembled aggregates. Appl Surf Sci. 2014;318:85–9.

53. Chatzieleftheriou S, Adendorff MR, Lagaros ND. Generalized potential energy finite elements for modeling molecular nanostructures. J Chem Inf Model. 2016;56:1963–78.

Modes of Cholesterol Binding in Membrane Proteins: A Joint Analysis of 73 Crystal Structures

Cong Wang, Arthur Ralko, Zhong Ren, Avia Rosenhouse-Dantsker, and Xiaojing Yang

Abstract Cholesterol is a highly asymmetric lipid molecule. As an essential constituent of the cell membrane, cholesterol plays important structural and signaling roles in various biological processes. The first high-resolution crystal structure of a transmembrane protein in complex with cholesterol was a human β_2-adrenergic receptor structure deposited to the Protein Data Bank in 2007. Since then, the number of the cholesterol-bound crystal structures has grown considerably providing an invaluable resource for obtaining insights into the structural characteristics of cholesterol binding. In this work, we examine the spatial and orientation distributions of cholesterol relative to the protein framework in a collection of 73 crystal structures of membrane proteins. To characterize the cholesterol-protein interactions, we apply singular value decomposition to an array of interatomic distances, which allows us to systematically assess the flexibility and variability of cholesterols in transmembrane proteins. Together, this joint analysis reveals the common characteristics among the observed cholesterol structures, thereby offering important guidelines for prediction and modification of potential cholesterol binding sites in transmembrane proteins.

Keywords Crystal structure · Membrane protein · Cholesterol-protein interactions · Singular value decomposition · Distance matrix

C. Wang · A. Ralko · Z. Ren · A. Rosenhouse-Dantsker
Department of Chemistry, University of Illinois at Chicago, Chicago, IL, USA

X. Yang (✉)
Department of Chemistry, University of Illinois at Chicago, Chicago, IL, USA

Department of Ophthalmology and Vision Sciences, University of Illinois at Chicago, Chicago, IL, USA
e-mail: xiaojing@uic.edu

© Springer Nature Switzerland AG 2019
A. Rosenhouse-Dantsker, A. N. Bukiya (eds.), *Direct Mechanisms in Cholesterol Modulation of Protein Function*, Advances in Experimental Medicine and Biology 1135, https://doi.org/10.1007/978-3-030-14265-0_4

Abbreviations

CARC   Inverted CRAC
CCM   Cholesterol consensus motif
CLR   Cholesterol
CRAC   Cholesterol recognition amino acid consensus
GPCR   G-protein coupled receptor
PDB   Protein Data Bank
RMSD   Root mean square deviation
SVD   Singular value decomposition
TM   Transmembrane
VDW   van del Waals

1   Introduction

Cholesterol is a lipid molecule characterized by a bulky steroid ring structure with a hydroxyl group on one end and a short hydrocarbon tail on the other (Fig. 1a). Compared to other membrane lipids with longer hydrocarbon tails such as phospholipids and glycolipids, cholesterol is a relatively rigid molecule owing to its four fused hydrocarbon rings. The overall dimension of a cholesterol molecule is about $19 \times 5.5 \times 4.7$ (Å), where its longest dimension roughly matches the thickness (~20 Å) of a lipid monolayer in cell membrane (Fig. 1b). Cholesterol has an asymmetric molecular shape with two methyl groups at C18 and C19 protruding towards the same side of the ring structure rendering a "rough" surface on one side and a "smooth surface" on the other (Fig. 1b). Depending on the bond distances from the C18 and C19 atoms, the atoms on the rim of the sterol ring form the asymmetric edges (denoted the "sharp" and "dull" edges) on the plane of the ring structure.

Cholesterol is one of the most important molecules in mammalian physiology [1–4]. In addition to its well-established structural roles in the integrity and mechanics of the cell membrane, cholesterol has emerged in recent years as a signaling molecule that modulates a wide range of signal transduction processes via direct interactions with membrane or membrane-associated proteins including receptors, transporters, ion channels and scaffold proteins [5–9]. However, the molecular mechanism of cholesterol-mediated signaling modulation has just begun to unravel. Thanks to the technological advances in structural biology studies of membrane proteins, the repertoire of membrane protein structures with bound cholesterol molecules has been significantly expanded in the Protein Data Bank (PDB) since the first high-resolution crystal structure of an engineered human β_2-adrenergic receptor in complex with cholesterol in 2007 [10, 11]. Such wealth of structural data offer many snapshots of the protein-cholesterol interactions at atomic resolution [12–36], which allow us to revisit a few well-known cholesterol binding motifs such as the CRAC [37, 38], CARC domains [39] and CCM [13].

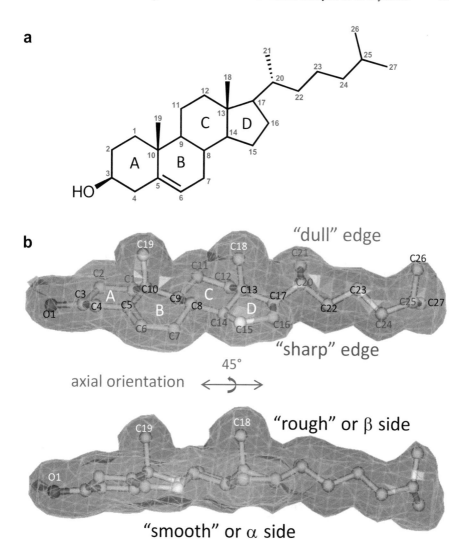

Fig. 1 The structure of cholesterol. (**a**) The chemical structure of a cholesterol molecule (PDB three-letter code: CLR). (**b**) The ball-and-stick model and electron density map of a bound CLR in a crystal structure of membrane protein (A2a adenosine receptor; PDBID: 4EIY) determined at 1.8 Å resolution. The top panel shows the naming convention of CLR where the atoms are labeled in different colors to highlight the molecular asymmetry of CLR. On one edge of the sterol ring, the C6/C7 and C15/C16 atoms render a thinner thus sharp edge than the other, where the C1/C3 and C11/C12 atoms together the C18/C19 methyl groups constitute a "dull" edge. The bottom panel shows CLR viewed from a different angle to illustrate the "smooth" or α side and "rough" or β side of CLR

In this work, we jointly analyze a collection of crystal structures that bind at least one cholesterol molecule. In the Protein Data Bank, cholesterol is represented by a three-letter code CLR, which is also used as a cholesterol abbreviation in this chapter. We examine the conformational flexibility of CLR manifested in the crystal structures of membrane proteins (Fig. 2). To survey the spatial distribution of CLR relative to the protein framework, we place a large number of CLR molecules in the same protein framework by superimposing 46 GPCR crystal structures with a reference structure (PDB ID: 4EIY), a high resolution (1.8 Å) structure of A2a adenosine receptor fusion protein that shows well-defined electron densities for three CLR molecules (Fig. 1b) [15]. We further analyze the modes of cholesterol binding by applying singular value decomposition (SVD) to an array of distance plots that uniquely characterize the CLR-protein interactions. Our joint analysis shows that the CLR binding is largely dictated by the shape complementarity along with the protein surface characteristics in transmembrane (TM) regions. While there is a clear consensus in the molecular orientation of the longest dimension of CLR, the axial orientation varies significantly as the CLR molecule of an asymmetric shape is docked in a niche pocket on the protein surface. We postulate that the bulky residues such as Phe and Trp are not only important for shaping the binding site but also for stabilizing the CLR binding via steric exclusion and ring stacking. Taken together, this analysis offers structural insights into various modes of direct interactions between CLR and membrane proteins.

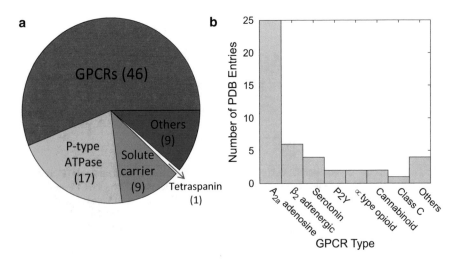

Fig. 2 Survey of cholesterol-bound protein structures in the Protein Data Bank. (**a**) Using 3 Å as a resolution cutoff, a total of 82 PDB structures are found to bind at least one cholesterol molecule identified by the three-letter code CLR. 73 out of 82 entries are membrane proteins. (**b**) Histogram of 46 GPCR structures under this study

2 Survey of Cholesterol-Bound Crystal Structures in the PDB

Using 3 Å as a resolution cutoff, we retrieved a total of 82 PDB entries that contain cholesterol as a ligand identified by a three-letter code CLR widely used for cholesterol. 73 out of 82 entries are annotated as membrane proteins, which fall into three major families: G-protein coupled receptors (GPCRs), P-type ATPase and solute carriers. For the purpose of investigating the CLR conformational flexibility and local protein-CLR interactions, we include 73 membrane protein structures. For illustration of the spatial distribution of CLR relative to the protein framework, we focus on 46 GPCR structures that share a conserved seven transmembrane helix (7TM) fold, which allow us to simultaneously examine a large number of CLR molecules under the same structural framework after alignment. Many protein chains have multiple CLR binding sites, and some PDB entries contain multiple protein chains in an asymmetric unit. To reduce redundancy and over-sampling, we only include one GPCR chain (usually chain A) along with all the associated CLR ligands in our analysis unless mentioned otherwise. Table 1 summarizes the PDB entries of GPCRs used in this study and each is annotated with the system name, diffraction resolution, crystal information including cell parameters and space group, IDs and membrane positioning for the CLR molecules together with the signaling state (activated/active or inactive) as reported. The table is assembled based on the information extracted from the PDB header and/or the paper cited by this entry.

3 Conformational Flexibility of CLR in Crystal Structures

To examine the conformational distribution of the CLR structures determined by X-ray crystallography, we extract 121 CLR molecules from 46 GPCR structures, including 26 PDB entries of the A2a adenosine receptor and 10 PDB entries of the β_2 adrenergic receptor (Table 1). We first generate a 2D distance matrix based on the atomic coordinates of each CLR molecule, which reduces a set of 3D coordinates to a 2D data matrix for further analysis (Fig. 3a). Each data point in the 2D matrix represents the inter-atomic distance between two corresponding atoms in CLR. Simply put, the farther apart the two atoms, the larger the distance value in such a distance matrix. This is similar to the distance geometry commonly used for protein structure determination by NMR [40]. To identify rigid and variable segments within the linear CLR molecule, we calculate the root mean square deviation (RMSD) matrix from a collection of distance matrices [41] (Fig. 3b). In a RMSD distance matrix, small variations in the interatomic distances identify a rigid body while the larger RMSD values reflects the greater variability that reveals the conformational flexibility between the rigid bodies. The RMSD matrix of 121 CLR structures from 46 PDB entries shows that the four hydrocarbon rings of CLR

Table 1 PDB entries for G-protein coupled receptor (GPCR) structures used in this study

Type PDBID	Reso. (Å)	Crystal info[a]	RMSD[b] (Å)	Signaling state[c]	CLR positioning in membrane[d] (CLR ID)	
A2a adenosine receptor		Space group: C2221			**Inner leaflet**	**Outer leaflet**
4EIY[e]	1.8	39.442 179.516 140.307 (Å)	0.00	Inactive		A2403 A2404 A2405
5 IU4	1.72	39.428 179.599 139.847 (Å)	0.287	Inactive		A2402 A2403 A2404 A2405
5 IU7	1.9	39.384 180.036 139.835 (Å)	0.228	Inactive		A2402 A2403 A2404 A2405
5 IU8	2	39.312 179.537 139.994 (Å)	0.294	Inactive		A2403 A2406 A2404 A2405
5IUA	2.2	39.560 179.979 139.924 (Å)	0.222	Inactive		A2402 A2403 A2404 A2405
5IUB	2.1	39.276 179.551 139.561 (Å)	0.368	Inactive		A2402 A2403 A2404
5JTB	2.8	39.648 179.793 139.810 (Å)	0.152	Inactive		A1203 A1202 A1204
5K2A	2.5	40.350 180.500 142.700 (Å)	0.345	Inactive		A1203 A1205 A1204
5K2B	2.5	40.360 180.740 142.800 (Å)	0.306	Inactive		A1203 A1205 A1204
5K2C	1.9	40.360 180.740 142.800 (Å)	0.346	Inactive		A1203 A1205 A1204
5K2D	1.9	40.360 180.740 142.800 (Å)	0.353	Inactive		A1203 A1205 A1204
5MZJ	2	39.246 179.867 139.398 (Å)	0.280	Inactive		A2402 A2403 A2404
5MZP	2.1	39.442 179.870 139.640 (Å)	0.322	Inactive		A2402 A2403 A2404 A2405

(continued)

Table 1 (continued)

Type PDBID	Reso. (Å)	Crystal info[a]	RMSD[b] (Å)	Signaling state[c]	CLR positioning in membrane[d] (CLR ID)	
5N2R	2.8	39.254 180.739 140.641 (Å)	0.394	Inactive		A2402 A2403 A2404
5NLX	2.14	40.330 180.070 142.660 (Å)	0.347	Inactive		A508 A507 A506
5NM2	1.95	39.428 179.599 139.847 (Å)	0.242	Inactive		A1224 A1223 A1222
5NM4	1.7	39.890 179.150 141.200 (Å)	0.286	Inactive		A509 A508 A510
5OLG	1.87	39.450 179.393 139.600(Å)	0.222	Inactive		A1203 A1202 A1205 A1204
5OLH	2.6	39.400 179.334 141.145 (Å)	0.353	Inactive		A1205 A1204 A1206
5OLV	2	39.429 180.774 140.903 (Å)	0.401	Inactive		A1203 A1205 A1204 A1206
5OLZ	1.9	39.369 179.247 140.066 (Å)	0.353	Inactive		A1203 A1202 A1205 A1204
5OM1	2.1	39.537 179.854 140.323 (Å)	0.320	Inactive		A1205 A1204 A1207 A1206
5OM4	2	39.465 179.109 140.032 (Å)	0.342	Inactive		A1203 A1205 A1204 A1206
5VRA	2.35	39.820 178.900 140.190 (Å)	0.142	Inactive		A2403 A2404 A2405
6AQF	2.51	39.837 180.973 140.574 (Å)	0.239	Inactive		A1203 A1202 A1204

(continued)

Table 1 (continued)

Type PDBID	Reso. (Å)	Crystal info[a]	RMSD[b] (Å)	Signaling state[c]	CLR positioning in membrane[d] (CLR ID)	
β₂-adrenergic receptor					**Inner leaflet**	**Outer leaflet**
2RH1	2.4	106.318, 169.240, 40.154 (Å) β = 105.62° (C2)	1.989	Inactive	A414 A413 A412	
3D4S	2.8	40.000, 75.700, 172.730 (Å) (P212121)	1.905	Inactive	A402 A403	
3NY8	2.84	40.711, 76.148, 174.207 (Å) (P212121)	1.671	Inactive	A1201 A1202	
3NY9	2.84	40.580, 75.900, 174.180 (Å) (P212121)	2.065	Inactive	A1201 A1202	
5D5A	2.48	107.000, 170.000, 40.500 (Å) β = 106.25° (C2)	1.933	Inactive	A1208 A1207 A1206	
5X7D	2.7	40.460, 75.710, 173.410 (Å) (P212121)	1.900	Inactive	A1203 A1204	
Serotonin receptor 2B					**Inner leaflet**	**Outer leaflet**
4IB4	2.7	60.571 119.750 170.607 (Å) (C2221)	2.865	Active	A2003	
4NC3	2.8	61.500 122.200 168.500 (Å) (C2221)	2.775	Active	A1203	
5TVN	2.9	59.195 119.177 170.990 (Å) (C2221)	3.597	Active	A2002	
6DRY	2.92	59.662 119.452 171.021 (Å) (C2221)	3.952	Active	A2002	
P2Y purinergic receptor						
4NTJ	2.62	98.650 156.430 47.770 (Å) β = 111.08° (C2)	7.236	Inactive	A1203	A1202
4PXZ	2.5	65.110 104.170 169.430 (Å) (C2221)	7.574	Active	A1202	
4XNV	2.2	66.270 66.270 239.070 (Å) (H3)	4.425	Inactive		A1102

(continued)

Table 1 (continued)

Type PDBID	Reso. (Å)	Crystal info[a]	RMSD[b] (Å)	Signaling state[c]	CLR positioning in membrane[d] (CLR ID)	
Cannabinoid receptor 1						
5XR8	2.95	66.830 73.610 139.640 (Å) (P21221)	4.12	Active	A1203	
5XRA	2.8	66.050 75.870 138.900 (Å) (P21221)	4.367	Active	A608	
μ-type opioid receptor						
5C1M	2.1	44.430 144.000 209.900 (Å) (I212121)	2.815	Active	A404	
4DKL	2.8	70.882 174.730 68.353 (Å) $\beta = 107.84°$ (C2)	3.015	Inactive		A614
Human chemokine CX3CL1 receptor						
4XT1	2.89	81.024 81.024 231.303 (Å) (I4)	4.095	Inactive		A401 A402
Glutamate receptor						
4OR2	2.8	67.361 86.552 168.277 (Å) (P212121)	9.19	Inactive	A1902 A1903 A1904 A1905 B1905 B1906	/
CC chemokine receptor type 9 (CCR9)						
5LWE	2.8	62.571 66.197 68.424 (Å) 74.02 64.72 62.29 (°) (P1)	3.410	Inactive	A417	
Endothelin receptor type B						
5X93	2.2	74.130, 147.480, 108.380 (C2221)	3.705	Inactive		A1213

[a]**Crystal info**: include the space group and cell parameters. However, the angles are not shown if they are implied by the space group

[b]**RMSD**: the root mean square difference between two aligned structures. The structural alignment and the associated RMSD values are generated using the *align* command in PyMol

[c]**Signaling state**: – Inactive: GPCR structures bound with an antagonist or inverse agonist – Active: GPCR structures bound with a natural substrate or agonist

[d]**CLR positioning in membrane**: annotates a cholesterol molecule (identified by the CLRID in a given PDB entry) by its association with the inner or outer leaflet of the membrane

[e]Reference PDB used for alignment of the GPCR structures

(denoted ring A, B, C and D, respectively) fall into two distinct rigid bodies defined by the internal RMSD values <0.2 Å (Fig. 1; Fig. 3b). Specifically, the rings A/B together with the C19 methyl group constitute one rigid body, while the rings C/D (atom C9-C18) form the other. The hydrocarbon tail shows a much higher mobility relative to the hydrocarbon rings. Not surprisingly, given the linear structure of CLR, the further apart the two segments, the larger the RMSD values (Fig. 3b).

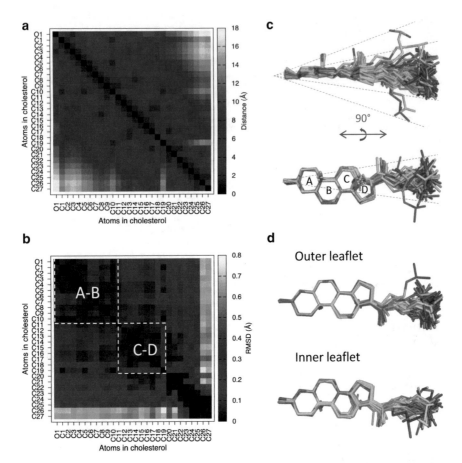

Fig. 3 Conformational flexibility of cholesterols in 46 GPCR structures. (**a**) The CLR conformation represented by a distance matrix, where the inter-atomic distances are color-coded at each grid point. (**b**) The root mean squares deviation (RMSD) distance matrix is calculated from the distance matrices of 121 CLR structures captured in 46 PDB entries. The identified rigid regions are highlighted by white dashed lines. (**c**) Structural alignment of 121 CLR molecules according to the atomic coordinates of rings A/B. CLR is colored from blue to red in spectrum according to the ascending order of the relative B-factors in each CLR. Grey dashed lines mark the extent of conformational variability in the tail group of CLR in two orthogonal views. (**d**) Comparisons between CLRs grouped by their membrane locations show that CLRs bound to the outer leaflet (top) generally have lower B-factors close to the head group while those bound to the inner leaflet show the opposite trend. CLR is viewed from the smooth side

To visualize the extent of conformational distribution in the linear structure of CLR, we further align these 121 CLR structures according to a subset of atomic coordinates from the rings A/B using the *align* command in PyMol based on least-squares fitting [42]. The alignment reveals significant displacements in the rings C/D and the hydrocarbon tail along the direction normal to the sterol rings although a majority of CLRs fall within an angular spread of about 25° (Fig. 3c). The hydrophobic tail of CLR shows conformational variability of a lesser extent in the plane defined by the sterol rings.

To address the CLR flexibility in crystal lattice, we also examined the atomic displacement parameters (also known as temperature or B-factor) of the CLR structures. In crystallography, B-factors measure the uncertainty in the atomic positions arising from thermal motions or flexibility in crystal lattice [43]. By coloring CLR according to the relative or normalized B-factor of each atom, we ask whether the CLR mobility bears any correlation with the binding mode and its protein environment. A majority of CLR molecules associated with membrane proteins in the outer leaflet show an ascending trend of B-factors from the hydroxyl head to the hydrocarbon tail (Fig. 3c) while those in the inner leaflet show an opposite trend (Fig. 3d), suggesting that the CLR mobility depends on the specific binding mode. It is plausible that the head becomes less flexible when the hydroxyl group of CLR forms hydrogen bonds with residues on the membrane surface while the sterol rings engage close van del Waals (VDW) interactions with the aromatic residues on the protein surface. As discussed in the following sections, both the rigid head and flexible tail of CLR are important structural features that facilitate CLR binding to various locations on the outer surface of membrane proteins as a structural and signaling lipid.

4   Spatial Distribution of CLRs in the GPCR Structures

To investigate the spatial distribution of CLR relative to the protein framework, we focus on the GPCR structures characterized by seven transmembrane (TM) helice, which enable us to examine CLRs under a common protein framework of the 7TM fold [44]. To address whether different GPCRs render different CLR binding sites, we align 46 GPCR structures along with the bound CLRs according to their protein backbones. With the PDB structure 4EIY as a reference [15], we used the structure-based sequence-independent protocol implemented in PyMol (i.e. *super* command) for structural alignment [42] (Table 1). We examined the properties or behaviors of the bound CLRs in the 7TM framework by coloring them by the GPCR type, absolute and relative B-factors, respectively (Fig. 4). An immediately notable consensus is that the CLR molecules always align the longest dimension with the normal direction of the membrane with its hydroxyl group anchored to the membrane surface via hydrogen bonds to a surface residue or neighboring lipid molecule. Interestingly, whether CLRs bind to the outer or inner leaflet of the membrane bilayers seems to bear some correlation with the GPCR type (Fig. 4a). Specifically, for A_{2a} adenosine receptor and cannabinoid receptors, CLRs favor the outer leaflet while for the β_2 adrenergic and serotonin receptors, they are more likely to be associated with the inner leaflet, at least in the crystal structures currently available in the PDB (Fig. 4a).

Another remarkable observation is that CLRs are not evenly distributed around the 7TM framework of GPCRs (Fig. 4). Three major binding sites are found. A prominent site is located near the kinked TM2 and TM3 helices (denoted TM2/3 site) that constitute a recessed surface between the TM1 and TM4 helices. A second

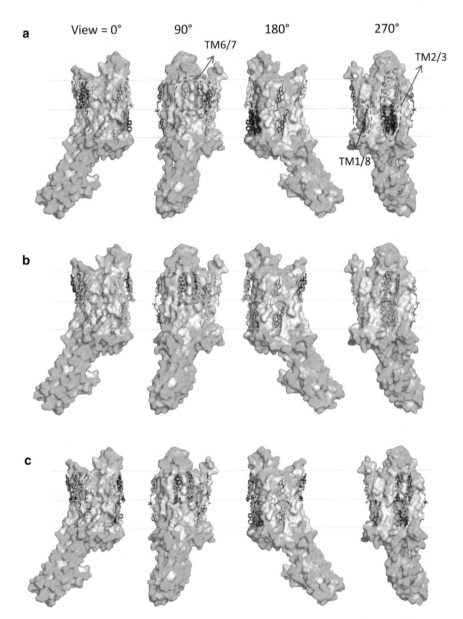

Fig. 4 Spatial distribution of bound cholesterol molecules in 46 GPCR structures. Structural alignment is based on the least-square fitting of the protein backbone structures using the high resolution crystal structure of A2a adenosine receptor (PDBID: 4EIY) as a reference. 4EIY also serves as a GPCR representative with a rendered surface colored according to hydrophobicity (green: polar; white: nonpolar), which is shown in four different views. In each panel, cholesterols (ball-and-stick) are colored according to either the GPCR type or B-factors. (**a**) Color by the GPCR type (A2aAR: yellow; β₂ adrenergic; blue; μ-type opioid: cyan; P2Y: warm pink; cannabinoid: forest; serotonin: light blue). See also Fig. 2b and Table 1. (**b**) Color by the B-factors relative to its protein framework according to the rainbow spectrum (ascending from blue to red). (**c**) Color by the relative B-factors in each CLR molecule using the rainbow spectrum color ascending from blue to red

site is a concave surface area at the outer membrane layer on the opposite side of the GPCR structure where the TM6 and TM7 helices sharply bend towards the protein interior near the middle segment (denoted TM6/7 site). The third site is a small niche pocket at the intersection of the TM1 and TM8 helices (denoted TM1/8 site) where CLRs align with the inner membrane segment of TM1 with the hydroxyl group anchored to TM8 via hydrogen bonds.

We identify some common structural features among these three sites. *First*, they all render recessed and elongated protein surfaces in the membrane segment that accommodate CLR, a rather rigid lipid molecule. *Second*, formation of these concave surfaces involves either two adjacent TM helices with the same bending points (TM2/3 and TM6/7) or at the crossing of two nearly perpendicular helices (TM1/8). *Third*, CLR is anchored to the membrane surface via hydrogen bonds mediated by its polar hydroxyl group. As a result, a single CLR molecule is associated with one membrane layer and seldom span across the bilayer. *Fourth*, CLRs seem to co-localize with bulky aromatic residues. Nearly all CLRs bound to the membrane proteins captured by crystallography show stacking or VDW interactions with at least one aromatic residue approaching from the smooth side. Such bulky residues may significantly alter the binding surface, which prevents the CLR binding due to steric hindrance. For example, a cluster of aromatic residues in the β_2 adrenergic receptor (Phe89, Trp99, Phe101, Phe104, Trp105 and Phe108) clearly prevent CLRs from binding to the outer membrane segment of the TM2/3 site, a major binding surface for two side-by-side CLR molecules in the A2a adenosine receptor structures [10, 11, 15] (Fig. 5a, c). The steric effect alone may account for the differential CLR binding between these two otherwise highly comparable GPCR structures (RMSD ~1.6 Å between 2RH1 and 4EIY) (Fig. 5) [10, 11, 15]. Compared to those associated with the inner leaflet, CLRs bound to the outer leaflet generally have lower B-factors relative to the protein framework evidenced by their cooler colors, suggesting a higher overall stability in the crystal lattice (Fig. 4b). However, the CLR disposition (inner vs. outer leaflet) does not seem to bear any correlation with the GPCR signaling state (inactive vs. activated) (Table 1).

5 Orientation Distribution of CLRs Relative to Membrane Proteins

CLR is a highly asymmetric molecule with distinct rough/smooth sides and sharp/dull edges (Fig. 1b). While CLR binding shows consensus in the longest dimension, their axial orientations vary significantly relative to the protein framework. To characterize the CLR-protein interactions, we search all residues within a 4.5-Å radius from any atoms in CLR as shown in the distance matrix (Fig. 6a). We then extract the shortest interatomic distance (Dmin) between each atom of CLR and the protein moiety where both the main chain and side chain atoms are considered in this calculation (Fig. 6b). We reason that this Dmin plot as a function of each CLR

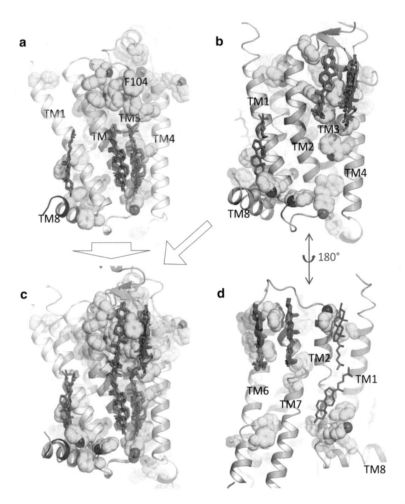

Fig. 5 Asymmetry distribution of cholesterols (CLR: blue sticks) and bulky residues (yellow spheres) in two types of GPCRs. (**a**) CLRs bind to the inner membrane leaflet in β_2 adrenergic receptor (PDB: 2RH1 in gray ribbon). (**b**) CLRs favor the outer membrane leaflet in A_{2a} adenosine receptor (4EIY in green ribbon). (**c**) Superposition of 2RH1 (grey) and 4EIY (green) structures along with their CLRs and aromatic residues (yellow spheres) show in the same view as (**a** and **b**). (**d**) The TM6/7 site of A_{2a} adenosine receptor is shown in a different view from (**b**) as indicated

atom not only captures the overall proximity between CLR and the protein matrix but also reveals the binding mode, that is, which side or edge of CLR engages in direct protein interactions. While the proximity informs on the binding affinity, the orientation describes the binding mode of CLR.

We therefore obtained the 178 Dmin plots using all CLR molecules in 73 membrane protein structures of different types (Fig. 6b). It is noteworthy that the Dmin standard deviations associated with the edge atoms (e.g. C6/C7, C11/C12, C15/C16) are larger than those for the centric atoms (i.e. C8/C9/C10, C13 and C17)

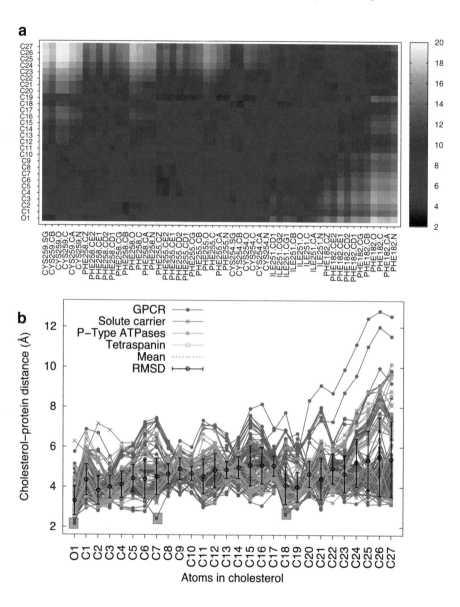

Fig. 6 Cholesterol-protein interactions. (**a**) Distance matrix between all atoms of a cholesterol (CLR) molecule and the protein atoms within 4.5 Å radius from CLR. (**b**) The shortest distance between each atom of CLR and its corresponding protein structure is plotted to characterize the interface at each CLR-binding site. All together, 178 cholesterol molecules from 73 PDB entries are included. Black error bars represent the standard deviations of the minimal protein-cholesterol distance for each atom. Shaded squares mark the three close contact points

apparently resulted from the variable CLR orientation. Furthermore, the top three shortest Dmin values are respectively associated with the O1 atom in the polar hydroxyl group, C7 on the sharp edge and C18 on the rough side, which represent the closest contact points between CLR and the protein matrix. Not surprisingly, the atoms forming the flexible tail group show the largest Dmin deviations.

To address whether CLR displays any preferred modes of binding in membrane proteins, we jointly analyze this collection of 178 Dmin plots by subjecting the corresponding data matrix to singular value decomposition analysis (Fig. 7a), from which we identify three significant components (Fig. 7b). In a SVD analysis, every decomposed SVD component is orthonormal to the other SVD components [41, 45]. In other words, the characteristic distance feature manifested in each left singular vector ($u1$, $u2$...) is unique and cannot be represented by any other components or their combination. By definition, the left singular vector ($u1$) corresponding to the top SVD component represents the average Dmin values between CLR and the protein matrix. The second component ($u2$) clearly singles out a structural feature uniquely associated with the flexible tail group. And the third component ($u3$) displays oscillating behaviors both in terms of sign and amplitude, supporting that $u3$ captures the axial orientation of CLR. The oscillation of $u3$ results from the CLR rotation about its longest dimension, (Fig. 1b), where the atoms on the same edge or side move together, therefore showing the concurrent or grouped changes in distance that occur as CLR spins (Fig. 7c, d).

To determine whether the flexibility ($u2$) and orientation ($u3$) of CLR bear any correlation with the CLR-protein distances ($u1$), we examine the three pairwise scatter plots between the right singular vectors associated with the decomposed components ($u1$, $u2$ and $u3$). Each dot in such a scatter plot corresponds to a bound CLR molecule from one of the 73 crystal structures examined. The $c1$-$c2$ scatter plot of the first two SVD components shows a wider distribution in the $c2$ dimension at larger $c1$ values suggesting that the tail group of CLR tends to be more flexible when the CLR is further away from the protein (Fig. 7c). The continuous distribution of c3 evident in both $c1$-$c3$ and $c2$-$c3$ plots suggest that CLR binding to the protein has no obvious preferred axial orientation. However, a positive correlation between c1 and c3 is observed with a slight skew towards the negative side in the $c3$ dimension. We postulate that CLR is more likely to approach the protein surface from its "sharp" edge (defined by C6-C7 and C15-C16), giving rise to the intimate CLR-protein interactions consistent with the shorter distances in $c1$.

6   Concluding Remarks

With the recent advances in structural biology of membrane proteins, the number of the atomic-resolution crystal structures with bound cholesterol molecules has increased significantly. This work presents a comprehensive survey of the direct cholesterol-protein interactions in 73 membrane protein structures in the Protein Data Bank. By examining the spatial and orientation distributions of cholesterol

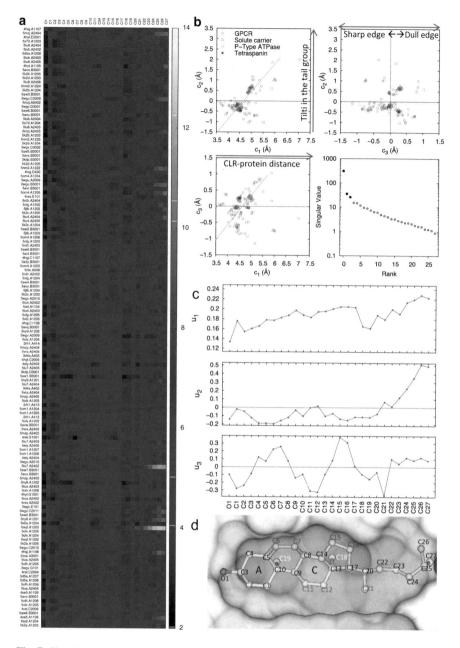

Fig. 7 Singular value decomposition (SVD) analysis of cholesterol-binding sites. (**a**) Input data matrix (178 × 28) used for SVD analysis. Each row (*PDBID.CLRID*) corresponds to one curve in Fig. 6b, representing the shortest distance (*D*min) between each atom of CLR and the corresponding protein chain in a given PDB entry. (**b**) Scatter plots between the top three singular components ranked by their significance as shown in the bottom-right panel. Each dot represents a *D*min distance plot colored by the protein type as labeled. Please note that the grey lines are not resulted from data fitting, they simply serve as a visual guide to aid the discussion. (**c**) The plots of the left singular vectors (*u1-u3*) for the top three components. (**d**) The ball-and-stick structure of a bound cholesterol molecule (4EIY CLR A2405) is shown with the rough side facing the protein surface in its concave binding site. It is intended to guide the interpretation of the decomposed distance plots in (**c**)

relative to the protein framework, we extract the common characteristics of cholesterol binding sites in membrane proteins. Our joint analysis shows that almost all observed cholesterol molecules orient the longest dimension along the normal direction within a membrane monolayer, which is guided by the hydroxyl group forming hydrogen bonds with residues on the membrane surface. Cholesterol clearly prefers the recessed protein surfaces and often engages in VDW interactions with aromatic residues via the smooth side of the sterol rings. However, despite its asymmetric shape, cholesterol does not seem to adopt specific axial orientations. The orientation of a loosely bound CLR near the protein surface is also influenced by interactions with the neighboring lipid or detergent molecules in crystal lattice. However, for those closely associated CLRs, where and how they bind to the membrane protein largely depend on the surface complementarity at individual locations, which is in turn dictated by the protein tertiary structure in a given functional state as well as the surface distribution of bulky residues. We must point out that many GPCR crystal structures were determined using engineered membrane proteins where one of the intracellular or extracellular loops are replaced by the well-behaved soluble proteins such as T4 lysozyme (T4L) or apocytochrome $b_{562}RIL$ (BRIL) [10, 11, 15]. Whether T4L or BRIL fusion has any effect on the mode of cholesterol binding remains to be seen. In the absence of a universal consensus sequence, this survey offers some guidelines for prediction and modification of a potential cholesterol site in membrane proteins.

References

1. Ikonen E. Cellular cholesterol trafficking and compartmentalization. Nat Rev Mol Cell Biol. 2008;9(2):125–38.
2. Yeagle PL. Modulation of membrane function by cholesterol. Biochimie. 1991;73(10):1303–10.
3. Yeagle PL. Cholesterol and the cell membrane. Biochim Biophys Acta. 1985;822(3–4):267–87.
4. Maxfield FR, van Meer G. Cholesterol, the central lipid of mammalian cells. Curr Opin Cell Biol. 2010;22(4):422–9.
5. Sheng R, et al. Cholesterol modulates cell signaling and protein networking by specifically interacting with PDZ domain-containing scaffold proteins. Nat Commun. 2012;3:1249.
6. Ramprasad OG, et al. Changes in cholesterol levels in the plasma membrane modulate cell signaling and regulate cell adhesion and migration on fibronectin. Cell Motil Cytoskeleton. 2007;64(3):199–216.
7. Goluszko P, Nowicki B. Membrane cholesterol: a crucial molecule affecting interactions of microbial pathogens with mammalian cells. Infect Immun. 2005;73(12):7791–6.
8. Gimpl G, Burger K, Fahrenholz F. Cholesterol as modulator of receptor function†. Biochemistry. 1997;36(36):10959–74.
9. Rosenhouse-Dantsker A, Mehta D, Levitan I. Regulation of ion channels by membrane lipids. In: Terjung R, editor. Comprehensive physiology. Hoboken, NJ: John Wiley & Sons, Inc.; 2012. https://doi.org/10.1002/cphy.c110001.
10. Cherezov V, et al. High-resolution crystal structure of an engineered human b2-adrenergic G protein–coupled receptor. Science. 2007;318(5854):1258–65.
11. Rosenbaum DM, et al. GPCR engineering yields high-resolution structural insights into β2-adrenergic receptor function. Science. 2007;318:9.

12. Morth JP, et al. Crystal structure of the sodium–potassium pump. Nature. 2007;450 (7172):1043–9.
13. Hanson MA, et al. A specific cholesterol binding site is established by the 2.8 Å structure of the human β2-adrenergic receptor. Structure. 2008;16(6):897–905.
14. Shinoda T, Ogawa H, Cornelius F, Toyoshima C. Crystal structure of the sodium–potassium pump at 2.4 Å resolution. Nature. 2009;459(7245):446–50.
15. Liu W, et al. Structural basis for allosteric regulation of GPCRs by sodium ions. Science. 2012;337(6091):232–6.
16. Wada T, et al. Crystal structure of the eukaryotic light-driven proton-pumping rhodopsin, acetabularia rhodopsin II, from marine alga. J Mol Biol. 2011;411(5):986–98.
17. Liu W, et al. Serial femtosecond crystallography of G protein–coupled receptors. Science. 2013;342:5.
18. Penmatsa A, Wang KH, Gouaux E. X-ray structure of dopamine transporter elucidates antidepressant mechanism. Nature. 2013;503(7474):85–90.
19. Wu H, et al. Structure of a class C GPCR metabotropic glutamate receptor 1 bound to an allosteric modulator. Science. 2014;344(6179):58–64.
20. Zhang K, et al. Structure of the human P2Y12 receptor in complex with an antithrombotic drug. Nature. 2014;509(7498):115–8.
21. Burg JS, et al. Structural basis for chemokine recognition and activation of a viral G protein-coupled receptor. Science. 2015;347(6226):1113–7.
22. Zhang D, et al. Two disparate ligand-binding sites in the human P2Y1 receptor. Nature. 2015;520(7547):317–21.
23. Penmatsa A, Wang KH, Gouaux E. X-ray structures of Drosophila dopamine transporter in complex with nisoxetine and reboxetine. Nat Struct Mol Biol. 2015;22(6):506–8.
24. Huang W, et al. Structural insights into μ-opioid receptor activation. Nature. 2015;524(7565):315–21.
25. Coleman JA, Green EM, Gouaux E. X-ray structures and mechanism of the human serotonin transporter. Nature. 2016;532(7599):334–9.
26. Chen Y, et al. Structure of the STRA6 receptor for retinol uptake. Science. 2016;353(6302):aad8266.
27. Zimmerman B, et al. Crystal structure of a full-length human tetraspanin reveals a cholesterol-binding pocket. Cell. 2016;167(4):1041–1051.e11.
28. Oswald C, et al. Intracellular allosteric antagonism of the CCR9 receptor. Nature. 2016;540(7633):462–5.
29. Martin-Garcia JM, et al. Serial millisecond crystallography of membrane and soluble protein microcrystals using synchrotron radiation. IUCrJ. 2017;4(4):439–54.
30. Cheng RKY, et al. Structures of human A 1 and A 2A adenosine receptors with xanthines reveal determinants of selectivity. Structure. 2017;25(8):1275–1285.e4.
31. Hua T, et al. Crystal structures of agonist-bound human cannabinoid receptor CB1. Nature. 2017;547(7664):468–71.
32. Shihoya W, et al. X-ray structures of endothelin ETB receptor bound to clinical antagonist bosentan and its analog. Nat Struct Mol Biol. 2017;24(9):758–64.
33. Johnson ZL, Chen J. ATP binding enables substrate release from multidrug resistance protein 1. Cell. 2018;172(1–2):81–89.e10.
34. Che T, et al. Structure of the nanobody-stabilized active state of the kappa opioid receptor. Cell. 2018;172(1–2):55–67.e15.
35. Jungnickel KEJ, Parker JL, Newstead S. Structural basis for amino acid transport by the CAT family of SLC7 transporters. Nat Commun. 2018;9(1):550. https://doi.org/10.1038/s41467-018-03066-6.
36. Zhang Z, Tóth B, Szollosi A, Chen J, Csanády L. Structure of a TRPM2 channel in complex with Ca2+ explains unique gating regulation. eLife. 2018;7:e36409. https://doi.org/10.7554/eLife.36409.

37. Epand RM. Cholesterol and the interaction of proteins with membrane domains. Prog Lipid Res. 2006;45(4):279–94.
38. Li H, Papadopoulos V. Peripheral-type benzodiazepine receptor function in cholesterol transport. Identification of a putative cholesterol recognition/interaction amino acid sequence and consensus pattern. Endocrinology. 1998;139(12):7.
39. Fantini J, Barrantes FJ. How cholesterol interacts with membrane proteins: an exploration of cholesterol-binding sites including CRAC, CARC, and tilted domains. Front Physiol. 2013;4:31. https://doi.org/10.3389/fphys.2013.00031.
40. Braun W. Distance geometry and related methods for protein structure determination from NMR data. Q Rev Biophys. 1987;19(3–4):115.
41. Ren Z. Molecular events during translocation and proofreading extracted from 200 static structures of DNA polymerase. Nucleic Acids Res. 2016;44(15):7457–74.
42. DeLano WL. Pymol: an open-source molecular graphics tool. CCP4 Newslet Protein Crystal. 2002;40:82–92.
43. Adams PD, et al. PHENIX: a comprehensive python-based system for macromolecular structure solution. Acta Crystallogr D Biol Crystallogr. 2010;66(2):213–21.
44. Kinoshita M, Okada T. Structural conservation among the rhodopsin-like and other G protein-coupled receptors. Sci Rep. 2015;5(1) https://doi.org/10.1038/srep09176.
45. Ren Z. Reverse engineering the cooperative machinery of human hemoglobin. PLoS One. 2013;8(11):e77363.

Part II
Cholesterol Binding Sites in Proteins: Case Studies

Effects of Cholesterol on GPCR Function: Insights from Computational and Experimental Studies

Sofia Kiriakidi, Antonios Kolocouris, George Liapakis, Saima Ikram, Serdar Durdagi, and Thomas Mavromoustakos

Abstract The extensive experimental and computational evidences revealed that cholesterol is involved in the drug binding to G protein-coupled receptor (GPCR) targets that is influenced by the membrane environment and external functions. These multifunctional factors make the understanding of the molecular mechanism of action in greater detail an entirely difficult task. Significant efforts have been made for better understanding the role of multi-directional specific, receptor-dependent interactions of cholesterol, and its effects on drug design and development. Additional efforts must be made in this complex system in order to shed more light on cholesterol molecular basis of action. The results of molecular simulations that complemented experimental data may reveal new aspects of GPCR-cholesterol interactions and may provide a comprehensive understanding of receptor function.

Keywords GPCRs · Cholesterol · Membrane bilayers · Cholesterol recognition interaction amino acid consensus (CRAC) · Multi-scale simulations

S. Kiriakidi · T. Mavromoustakos (✉)
Laboratory of Organic Chemistry, Department of Chemistry, National and Kapodistrian University of Athens, Athens, Greece
e-mail: tmavrom@chem.uoa.gr

A. Kolocouris
Section of Pharmaceutical Chemistry, Department of Pharmacy, School of Health Sciences, National and Kapodistrian University of Athens, Athens, Greece

G. Liapakis
Department of Pharmacology, School of Medicine, University of Crete, Heraklion, Crete, Greece

S. Ikram · S. Durdagi (✉)
Computational Biology and Molecular Simulations Laboratory, Department of Biophysics, School of Medicine, Bahcesehir University, Istanbul, Turkey
e-mail: serdar.durdagi@med.bau.edu.tr

© Springer Nature Switzerland AG 2019
A. Rosenhouse-Dantsker, A. N. Bukiya (eds.), *Direct Mechanisms in Cholesterol Modulation of Protein Function*, Advances in Experimental Medicine and Biology 1135, https://doi.org/10.1007/978-3-030-14265-0_5

1 Introduction

Cholesterol is a well-known lipid molecule that exists in the membranes of the most animal species. In recent years, the roles of the molecular components that constitute membranes, such as cholesterol, have been emphasized in order to understand their biological role. Cholesterol is not only a membrane component that affects drug diffusion but also interacts with physiological proteins. These proteins include G protein-coupled receptors (GPCRs) in which cholesterol has specific binding sites. Based on proteins known to interact with cholesterol, a cholesterol recognition/interaction amino acid consensus (CRAC) has been suggested. Besides, it is well known that high concentrations of cholesterol in the atherosclerotic plaques are leading to heart diseases. Deviations in cholesterol homeostasis contribute not only to heart diseases and stroke, but also to common sporadic and complex disorders, including type II diabetes and Alzheimer's disease. Thus, cholesterol is considered to play a dual role in the human health. From one aspect, exerts many physiological functions but on the other hand, if physiological concentrations are exceeded, it can be detrimental to the human health [1, 2].

Scheme 1 illustrates the possible interactions of cholesterol and the factors affecting its functions on GPCRs. The following sections summarize some of the effects of cholesterol on membranes and discuss the recent evidence of modulation of their interactions and functions with GPCRs. Recent reviews and the special issue of the Chemistry and Physics on cholesterol (2016) are excellent sources to provide more information on properties and functions of cholesterol [3–6]. The purpose of this review paper is to formulate the effects of cholesterol on GPCRs and its implications on human health with some suggestions to promote prospective research activities.

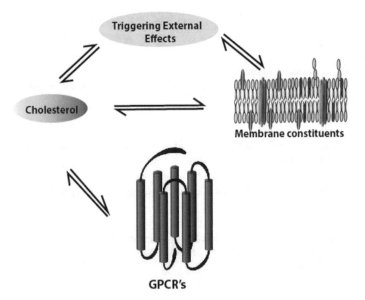

Scheme 1 Cholesterol is a major membrane component and interacts with membrane constituents and GPCRs. Its effects are also governed by external factors. The proper equilibrium of all these factors results in cholesterol exerting its beneficial results. However, in pathological states cholesterol may cause equally detrimental effects

## 2	3D Structure of Cholesterol

The biologically vital function of cholesterol is due to its unique molecular features (Fig. 1). Its structure consists of four linked hydrocarbon rings forming the bulky steroid structure. Cholesterol as shown in Fig. 1 is a sterol i.e. combines the skeleton of a steroid with a hydroxyl group. Cholesterol possesses two orthogonal methyl groups on its quaternary centers, i.e. C18 and C19 methyl groups, pointing to the β-phase and creating the so called "β rough surface". It lacks such methyl groups at the two tertiary carbons C8 and C14 thus creating the so called "α smooth surface". It contains a flexible isooctyl hydrocarbon tail linked to the carbon (C17) of the cyclopentyl ring D and a 3β-hydroxyl (head group) linked to the other end of the A ring. This hydroxyl polar group provides the amphipathicity to the lipophilic molecule. Thus, cholesterol contains mostly a structural rigid part, a flexible chain and a hydroxyl group fixed at the 3β position. This 3β-hydroxyl group is considered as the anchor for its position in the vicinity of the lipid-water interface, while the rest of the hydrophobic core aims to fit between the hydrophobic chains of the lipids. Another important structural feature of cholesterol is its chirality. Cholesterol is a chiral molecule with multiple chiral centers [7, 8].

## 3	Cholesterol Rafts

As cholesterol contains only a single polar group it is more hydrophobic than a phospholipid. Consequently, although most cholesterol molecules in a membrane are located with their hydroxyl groups in the headgroup region of the lipid bilayer and with their hydrophobic ring systems almost perpendicular to the plane of the membrane, a proportion of the cholesterol molecules occupies positions deep within the fatty acyl chain region of the bilayer, close to the bilayer center, as shown by neutron diffraction studies and all atoms (AA) and coarse-grained (CG) molecular dynamics (MD) simulations. The latter position is consistent with the reported high rate of flip-flop of cholesterol across the membrane [9–14].

Ceramide is a subfamily of sphingolipid and like cholesterol contains a highly hydrophobic carbon skeleton linked to a hydroxyl group. Therefore, it is directed in a similar manner in the bilayer leading to a natural displacement or exchange of cholesterol [15].

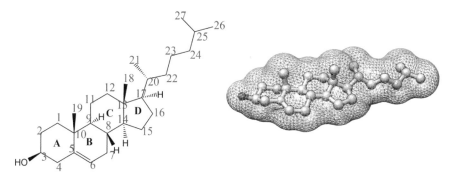

Fig. 1 2D and 3D molecular structures of cholesterol

Ceramide can be generated enzymatically from sphingomyelin that interacts with cholesterol for lipid raft formation. Consequently, cholesterol by enriching lipid rafts, serves as platform for signaling transduction of proteins in the plasma membrane. On the other hand, the solubility of cholesterol in the membrane is limited as it produces crystals in higher molar fractions and it is responsible for modifying physical properties of lipid bilayers [16]. An important structural feature of the binding of sterol molecules to water-soluble sterol-binding proteins is the hydrophobicity of the sterol ring system. However, together with hydrophobic effects, cholesterol-membrane protein interactions are driven predominantly by hydrogen bonding interactions formed by the cholesterol hydroxyl group. Indeed, all resolved structures for membrane protein-bound cholesterol depict the presence of such hydrogen bonds [14].

4   Cholesterol: Drug Interactions on GPCRs

The focus of this review is to highlight the recent findings about cholesterol binding to the GPCRs. GPCRs constitute the largest superfamily in the eukaryotic cells. They are characterized by a highly conserved seven transmembrane (TM) core architecture interconnected by extracellular and intracellular loops. They are classi-fied into five families. GPCRs are fundamentally important as versatile and dynamic in the signal transduction and cellular response to different kinds of extracellular stimuli. They adopt various active and inactive conformations that are stabilized by appropriate ligands acting as agonists or antagonists. They also represent major i.e. ~30% drug targets in all clinical areas [17, 18].

The abundant cholesterol in the plasma membrane of eukaryotic cells is close to all integral membrane proteins, and the function of some GPCRs has been shown to be dependent on cholesterol [19, 20].

The fluidity and curvature of the membrane, lateral pressure and membrane thickness can influence the cholesterol approach, as well as itself and other drugs binding to the GPCRs [20].

Cholesterol may approach to the GPCR-membrane environment with direct or indirect pathways (Fig. 2). It affects the incorporation of drug molecules in the receptor-binding site if they are considered to act through a membrane pathway. For example, it was found that due to the denser lipid packing in the cholesterol rich lipid raft, losartan, a drug molecule that acts on the AT1 receptor from a GPCR fam-ily, is likely to be excluded from this area, and preferentially found in the more fluid plasma membrane regions [21, 22]. In this region, losartan can accumulate and finally reach the AT1 receptor site (Fig. 3).

5   Direct Action of Cholesterol on GPCR Receptors

The following observations in a recent review article [3] are of great importance: (a) cholesterol can form hydrogen bonds through its 3β-hydroxyl group with the TM surfaces of GPCRs including exposed polypeptide backbone bonds and exposed

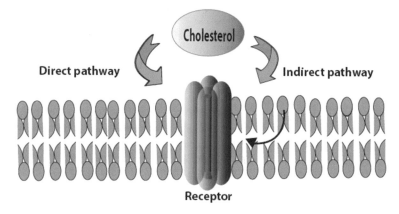

Fig. 2 Indirect or two step mechanism of action for cholesterol. In the first step cholesterol is postulated to embed itself in the lipid matrix and in the second step is laterally diffused to the active site of GPCR. In the direct mechanism it reaches the active site of the receptor through its mouth

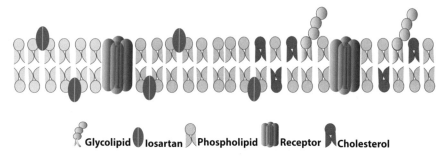

Fig. 3 A possible mode of action for losartan where cholesterol interferes with its action. Losartan avoids the cholesterol lipid rafts and localizes itself in other regions where it can easily diffuse and reaches the AT1 receptor site

side chain polar residues; (b) these hydrogen bonding interactions depend on the partitioning of cholesterol between surface and deep locations in the lipid bilayer, and on steric interactions between the cholesterol molecules and the protein surface and lipid fatty acyl chains close to the binding site; (c) changes in entropy can be important as the binding, for example in a deep cleft, decreases the entropy of the receptor complex but consequently increases the entropy for the phospholipid fatty acyl chains; (d) hydrogen bonds are observed in all high affinity GPCR-cholesterol complexes but also many non-hydrogen bonded backbone carbonyl groups and side chains exposed on the TM surface of the GPCRs do interact with cholesterol according to the MD simulations; (e) multiple binding sites of cholesterol in the membrane interface and in the deep regions make difficult the direct experimental evidence for revealing the importance of deep cholesterol sites, i.e., with GPCRs. (f) The presence of cholesterol or cholesterol hemisuccinate increases the thermal stability of many GPCRs in detergent micelles and in lipid bilayers and, in some cases, increases the affinity for agonists with unclear mechanisms; (g) addition of cholesterol to the β2 adrenergic receptor expressed in Sf9 insect cells, where the cholesterol level is low, led to a two-fold increase in affinity for the partial inverse agonist timolol but no change for the full agonist isoproterenol; (h) the binding interaction of deep cholesterol molecules with the A_{2A} receptor is comparable to that with the β2 adrenergic receptor; (i) serotonin$_{1A}$ receptor was the first one found to exhibit cholesterol-dependent functional modulation in terms of both ligand binding and G-protein coupling [18]. The serotonin$_{1A}$ receptor is an important neurotransmitter receptor of the GPCR superfamily and is implicated in the generation and modulation of various cognitive, behavioral and developmental functions. By analyzing the maximum occupancy of cholesterol molecules at different sites on the serotonin$_{1A}$ receptor, several cholesterol hot-spots have been identified [19]. These hot-spots correspond to the sites of maximum occupancy i.e. which was on average higher than the remaining regions of the receptor. The cholesterol occupancy sites were observed to be present on both near the plasma and deep in the TM region [23].

Analysis of sensitivity of the receptor in conditions of thermal deactivation, pH, and proteolytic digestion in control has been performed. Cholesterol-depletion with methyl-β-cyclodextrin (Mβ CD), a water-soluble polymer with a non-polar central cavity that has been shown to selectively and efficiently extract cholesterol from cellular membranes by incorporating it in a central nonpolar cavity, in cholesterol-enriched membranes, comprehensively demonstrates that membrane cholesterol stabilizes the serotonin$_{1A}$ receptor [24] (Fig. 4). Cholesterol depletion has been shown to decrease GnRH-mediated activation of extracellular signal-related kinase (ERK) and c-Fos gene induction. Repletion of membrane cholesterol rescued raft localization and GnRH receptor signaling to ERK and c-Fos [25]. Similarly, the Neurokinin type 1 receptor-mediated intracellular signaling that induces phosphorylation and activation of the extracellular signal-regulated kinase (ERK) was abolished after cholesterol depletion with Mβ CD [26].

Cholesterol replenishment of solubilized membranes to explore the stereospecific stringency of cholesterol for receptor function was studied using two stereoisomers of cholesterol, *ent*-cholesterol (an enantiomer of cholesterol) and *epi*-cholesterol (a diastereomer of cholesterol). It has been shown that while *ent*-cholesterol could replace cholesterol in supporting receptor function, *epi*-cholesterol could not. These results imply that the requirement of membrane cholesterol for the serotonin$_{1A}$ receptor function is diastereospecific, yet not enantiospecific (Fig. 5) [27].

Homology modeling of serotonin$_{1A}$ receptor revealed that ligands exhibit lower binding energies when docked to the receptor in the presence of cholesterol, thereby implying that membrane cholesterol facilitates ligand binding to the serotonin$_{1A}$ receptor [28, 29].

Today, it is estimated that there are more than 50 receptors exhibiting cholesterol-dependent function. A crystal structure of β2-receptor with molecules of cholesterol is shown in Fig. 6. X-ray diffraction data demonstrates that GPCRs exhibit cholesterol-dependent conformational dynamics and the presence of high affinity binding sites on cholesterol. As we discussed above, cholesterol is influenced by lipids, i.e. phospholipids and sphingolipids, which can accommodate ligand binding by GPCRs.

Cholesterol Defficient Membrane: unstable receptor

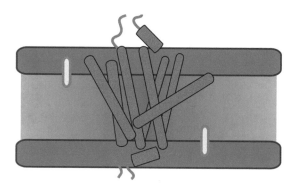

Cholesterol Rich Membrane: stable receptor

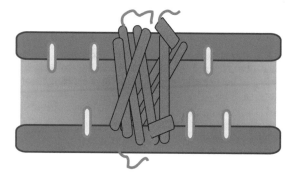

The Nobel Prize laureate in 2012 Brian Kobilka and his co-workers resolved two cholesterol molecules at the groove formed by TM helices I, II, III and IV (Fig. 7). This site in the crystal structure of the β2-adrenergic receptor is well known as Cholesterol Consensus Motif (CCM) [30].

Recently, these experimental data and also those obtained with serotonin receptor have been compared with AA and CG MD simulations. The motif on TM helix V was found as one of the sites with high cholesterol occupancy, thereby confirming its role as a putative cholesterol binding motif. These computational results that complemented experimental data, revealed new aspects of GPCR-lipid interactions and provided a comprehensive understanding of receptor function and made predictions that can be tested in the future [5].

The adenosine A$_{2A}$ receptor (A$_{2A}$R) is a class-A GPCR that plays a major role in the heart and brain by regulating oxygen consumption and blood flow. In fact, in the central nervous system, the A$_{2A}$R constitutes a potential therapeutic target for the treatment of Alzheimer and Parkinson's disease. Gonzalez et al. confirmed the presence of cholesterol inside the receptor by chemical modification of the A$_{2A}$R interior

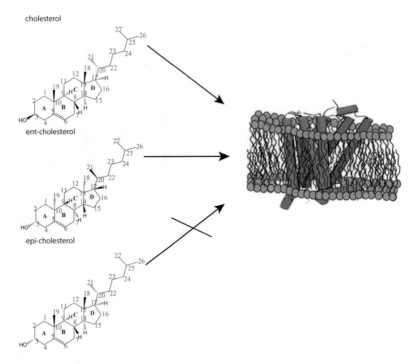

cholesterol

ent-cholesterol

epi-cholesterol

Fig. 5 A schematic representation of the reconstituted serotonin$_{1A}$ receptor replenished with ent- and epi- cholesterols. Replenishment with cholesterol and *ent*-cholesterol supports the function of the receptor, but not with the replenishment of *epi*-cholesterol

Fig. 6 Direct interactions of cholesterol with β2-adrenergic receptor. Intracellular part of the receptor is shown on the top and extracellular loops are at the bottom

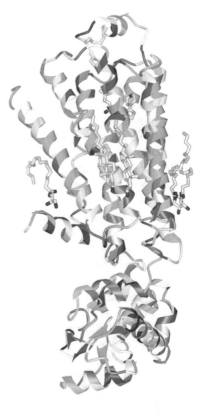

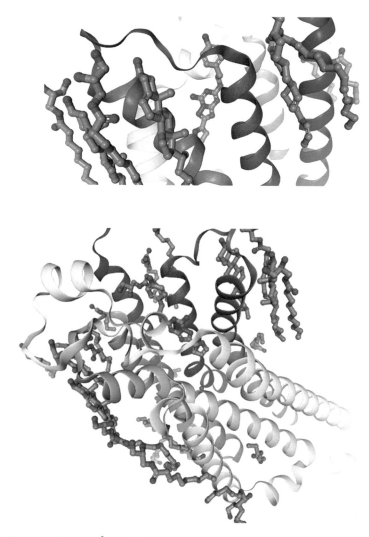

Fig. 7 Human $A_{2A}R$ at 1.9 Å resolution (PDB 5K2D) using X-ray free-electron lasers (XFELs) with the ligand agonist ZM241385. A cholesterol is also crystallized in the vicinity of the $A_{2A}R$ agonist

using a biotinylation assay. Combined long-scale MD simulations and experimental results showed that cholesterol can compete with orthosteric ligands by entering the receptor interior from the membrane side. Thus, Gonzalez et al. showed that cholesterol's impact on $A_{2A}R$-binding affinity goes beyond pure allosteric modulation and unveils a new interaction mode between cholesterol and the $A_{2A}R$ that could potentially apply to other GPCRs [6].

Cholesterol binding to the $A_{2A}R$ at allosteric sites has been previously demonstrated by the high-resolution X-ray crystal structures (PDB ID 3EML) [31] and PDB ID 5K2D [32]. Computational work has further quantified allosteric cholesterol binding to the receptor surface [33, 34] and it suggests a stabilizing effect on the apo-form of the $A_{2A}R$ [33]. However, the ability of cholesterol to impact ligand-binding properties through allosteric modulation at the $A_{2A}R$ remains unclear. Two new cholesterol interaction sites on the $A_{2A}R$ have been detected [35].

The home message from all the experimental and in silico studies of GPCRs with cholesterol is that it can affect various specific receptor interactions such as between its sterol ring and isooctyl segments with hydrophobic amino acids (i.e., Leu, Ala and Val), between C5–C6 double bond and the aromatic amino acids (i.e., Tyr, Phe), and between 3β-hydroxyl and the polar amino acids like Arg etc. [36].

6   Cholesterol May Regulate the Conformational Dynamics of the Receptor

Three conformational states of the receptor, i.e. the inactive state, active state, and the active state with a mini-G_S protein bound were simulated to study the impact of protein-lipid interactions on the receptor activation [37]. The simulations studies have revealed that three specific lipids i.e. glycolipid GM3, cholesterol and phosphatidylinositol 4,5-bisphosphate (PIP2), form stable interactions with the receptor, differentiating these from bulk lipids such as phosphatidylserine (PS), phosphatidylethanolamine (PE) and phosphatidylcholine (PC). In total, nine specific lipid-binding sites were revealed. The strength of lipid interaction with these sites depends on the conformational state of the receptor, suggesting that these lipids may regulate the conformational dynamics of the receptor. These results indicate likely allosteric effects of bound lipids in regulating the functional behavior of GPCRs, providing a springboard for design of allosteric modulators of these biomedically important receptors [37].

7   GPCR Oligomerization

There is growing evidence of GPCRs oligomerization which is essential for different functions. It is believed to be an important determinant for GPCR function and cellular signaling and implicated in proper folding of receptors proteins. Therefore, oligomerization structure provides the framework for efficient and controlled signal transduction and also a target for rational drug design. The role of membrane cholesterol and of the actin cytoskeleton in GPCR oligomerization, was revealed using a combined approach of homo-fluorescence resonance energy transfer (FRET) and CG MD simulations [33].

MD simulations aided to pinpoint TM helices involved in forming the receptor dimer interfaces [38–42]. The receptor dimer interface appears to be dependent on membrane cholesterol content. In addition, the possibility of homo- and hetero-oligomerization of GPCRs provides diversity and potential applications to future drug discovery [5]. Compounds that may prefer either GPCR dimers or monomers are under consideration for novel drug development. Receptor oligomerization state under various pathophysiological conditions can give rise to pharmacological diversity and opens new avenues for therapeutics. For example, it can lead to the development of drugs that inhibit the activity of heterodimers of angiotensin II receptor type 1 (AT1R) and chemokine type 2 receptor (CCR2) for the treatment of chronic kidney disease. Post-synaptic heterodimers of $A_{2A}R$ and dopamine D2 receptor can be the target of $A_{2A}R$ antagonists in the treatment of Parkinson's disease.

The presence of constitutive oligomers of the serotonin$_{1A}$ receptor has been proposed. Cholesterol mediates effects in GPCR dimers which appear to be receptor-specific as found in β2-adrenergic serotonin$_{1A}$ heterodimers. The most favorable cholesterol interaction site in the $β_2$-adrenergic receptor was at TM helix IV, as opposed to several cholesterol interaction sites of comparable occupancies, including TM helices I, V and VI in the serotonin$_{1A}$ receptor. This leads to the destabilization of the TM helix IV/V interface in the β2-adrenergic receptor in the presence of cholesterol, and an opposing stabilization of the flexible helix I-II interface in the dimer regime in the serotonin$_{1A}$ receptor [43–45].

Certain GPCRs can be expressed and function in *Escherichia coli* membranes, which lack cholesterol [46].

In a commentary manuscript, the significance of palmitoylation and dimer formation was pointed out [47]. For example, in the structure of β2AR, cholesterol was palmitoylated, contributing also in GPCR dimer formation. It is possible that this effect was largely observed due to crystal packing conditions but such an arrangement could influence GPCR activity in vivo. In contrast to the β2AR crystal structure, the palmitoate group on μ-opioid receptor (OPRM1) is not located on the carboxy-terminal tail but on the intracellular side of TM III helix. In $β_2$AR, the dimer interface involves TM1 and helix 8 in the carboxy-terminal tail whereas the OPRM1 dimer interface is predicted to be between TM4 of each protomer, with the palmitoate bound to the carboxy terminal side of TM3. This difference in receptor interface may be driven by the location of palmitoylation and it is possible that regulation of palmitoylation states could dynamically influence the dimerization interfaces of GPCRs. The complex interplay between receptor, cholesterol and palmitoate [48] lends support to the model suggested by the crystal structure of $β_2$AR by demonstrating a role for sterols and lipids in GPCR dimerization in vivo [49].

Insights in the GPCRs oligomerization states are outlining the effectiveness of CG MD simulations and the results obtained using Martini force field. It is interesting the optimistic notion mentioned in this recent review article: "the nature of the data produced by this method will increase in complexity to become closer and closer to realistic membrane compositions with time scales reaching experimental observables and it is incredibly exciting to envision the data that will be accessible in 5–10 years" [50].

Modern computational resources allow further refining of structural data and deepen our understanding of cholesterol-GPCR interactions. Computational chemistry offers a range of simulation, multi-scale modeling and virtual screening tools for definition and analysis of protein-cholesterol and membrane-cholesterol interactions. Development of new techniques on statistical methods and free energy simulations helps for better understanding of these chemical interactions.

8 Conclusions

Cholesterol is an amphoteric molecule associated with life and health. [51] The long steroid skeletal and isooctyl alkyl chain in cholesterol molecule add to its high lipophilic character while the 3β-hydroxyl group contributes to its polarity and amphiphilicity. The β surface of cholesterol contains two methyl groups and it is rough while the α surface does not contain substitutions and is therefore a smooth surface. These structural characteristics along with its high chirality are responsible for its versatile interactions with the lipid components of lipid bilayers as well as proteins. Among the proteins that interact with cholesterol are the GPCRs, a receptor group that is implicated in pharmacological function and drug discovery.

So far, AA [4, 34], CG MD simulations [4, 49, 50] and experimental results from X-ray crystallography [30], solid state NMR [46], ESR [52] and FRET [53], clearly show primary binding sites of cholesterol and its ability to modulate the function of GPCRs and interference with the drug action. MS-spectroscopy also seems to be a promising method allowing the detection of stable binding positions of lipids to membrane proteins [54, 55].

With the development of high throughput techniques, it will further be possible to understand the intricate mechanism of cholesterol and membrane mediated GPCRs activation which will ultimately pave a way for the development of novel therapeutic approaches.

References

1. Grouleff J, Sheeba JI, Katrine KS, SchiØtt B. The influence of cholesterol on membrane protein structure, function, and dynamics studied by molecular dynamics simulations. Biochim Biophys Acta. 2015;1848:1783–95.
2. Zhu C, Ichetovkin M, Kurz M, Zycband E, Kawka D, Woods J, He X, Plump A, Hailman E. Cholesterol in human atherosclerotic plaque is a marker for underlying disease state and plaque vulnerability. Lipids Health Dis. 2010;9:61.
3. Chattopadhyay A, Epand RM, editors. Properties and functions of cholesterol. Chem Phys Lipids. 2016;199:1–186.
4. Genheden S, Jonathan WE, Lee AG. G protein coupled receptor interactions with cholesterol deep in the membrane. Biochim Biophys Acta. 2017;1859:268–81.
5. Durba S, Xavier P, Mohole M, Chattopadhyay A. Exploring GPCR−lipid interactions by molecular dynamics simulations: excitements, challenges, and the way forward. J Phys Chem B. 2018;122:5727–37.
6. Guixà-González R, Albasanz JL, Rodriguez-Espigares I, Pastor M, Sanz F, Martí-Solano M, Manna M, Martinez-Seara H, Hildebrand PW, Martín M, Selent J. Membrane cholesterol access into a G-protein-coupled receptor. Nat Commun. 2017;8:1–12.

7. Dahl JS, Dahl CE, Bloch K. Sterol in membranes: growth characteristics and membrane properties of Mycoplasma capricolum cultured on cholesterol and lanosterol. Biochemistry. 1980;19:1467–72.

8. Marquardta D, Kŭcerkac N, Wassalle SR, Harrounf TA, Katsaras J. Cholesterol's location in lipid bilayers. Chem Phys Lipids. 2016;199:17–25.

9. Genheden G, Essex JW, Lee AG. G protein coupled receptor interactions with cholesterol deep in the membrane. Biochim Biophys Acta. 2017;1859:268–81.

10. Harroun TA, Katsaras J, Wassall SR. Cholesterol hydroxyl group is found to reside in the center of a polyunsaturated lipid membrane. Biochemistry. 2006;45:1227–33.

11. Marrink SJ, de Vries AH, Harroun TA, Katsaras J, Wassall SR. Cholesterol shows preference for the interior of polyunsaturated lipid membranes. J Am Chem Soc. 2008;130:10–1.

12. Kucerka N, Perlmutter JD, Pan J, Tristram-Nagle S, Katsaras J, Sachs JN. The effect of cholesterol on short- and long-chain monounsaturated lipid bilayers as determined by molecular dynamics simulations and x-ray scattering. Biophys J. 2008;95:2792–805.

13. Bennett WFD, MacCallum JL, Hinner MJ, Marrink SJ, Tieleman DP. Molecular view of cholesterol flip-flop and chemical potential in different membrane environments. J Am Chem Soc. 2009;131:12714–20.

14. Song YL, Kenworthy AK, Sanders CR. Cholesterol as a co-solvent and a ligand for membrane proteins. Protein Sci. 2014;23:1–22.

15. García-Aribas AB, Alonso A, Goñi FM. Cholesterol interactions with ceramide and sphingomyelin. Chem Phys Lipids. 2016;199:26–34.

16. Brzustowicz MR, Cherezov V, Zerouga M, Caffrey M, Stillwell W, Wassall SR. Controlling membrane cholesterol content. A role for polyunsaturated (docosahexaenoate) phospholipids. Biochemistry. 2002;41:12509–19.

17. Rosenbaum DM, Rasmussen SGF, Kobilka BK. The structure and function of G-protein-coupled receptors. Nature. 2009;459:356–63.

18. Sengupta D, Prasanna X, Madhura M, Chattopadhyay A. Exploring GPCR−lipid interactions by molecular dynamics simulations: excitements, challenges, and the way forward. J Phys Chem B. 2018;122:5727–37.

19. Sengupta D, Chattopadhyay A. Identification of cholesterol binding sites in the 645 serotonin1A receptor. J Phys Chem B. 2012;116:12991–6.

20. Gimpl G. Interaction of G protein coupled receptors and cholesterol. Chem Phys Lipids. 2016;199:61–73.

21. Kellici TF, Tzakos AG, Mavromoustakos T. Rational design and synthesis of molecules targeting the angiotensin II Type 1 and Type 2 receptors. Molecules. 2015;20:3868–97.

22. Hodzic A, Zoumpoulakis P, Pabst G, Mavromoustakos T, Rappolt M. Losartan's affinity to fluid bilayers couples to lipid/cholesterol interactions. Phys Chem Chem Phys. 2012;14:4780–8.

23. Sengupta D, Chattopadhyay A. Molecular dynamics simulations of GPCR−cholesterol interaction: an emerging paradigm. Biochim Biophys Acta. 2015;1848:1775–82.

24. Roopali S, Chattopadhyay A. Membrane cholesterol stabilizes the human serotonin1A receptor. Biochim Biophys Acta. 2012:2936–42.

25. Navratil AM, Bliss SP, Berghorn KA, Haughian JM, Farmerie TA, Graham JK, Clay CM, Roberson MS. Constitutive localization of the gonadotropin-releasing hormone (GnRH) receptor to low density membrane microdomains is necessary for GnRH signaling to ERK. J Biol Chem. 2003;278:31593–602.

26. Monastyrskaya K, Hostettler A, Buergi S, Draeger A. The NK1 receptor localizes to the plasma membrane microdomains, and its activation is dependent on lipid raft integrity. J Biol Chem. 2005;280:7135–46.

27. Jafurulla MD, Bhagyashree DR, Sugunan S, Jean-Marie R, Covey DF, Chattopadhyaya A. Stereospecific requirement of cholesterol in the function of the serotonin1A receptor. Biochim Biophys Acta. 2014;1838:158–63.

28. Paila YD, Shrish T, Sengupta D, Chattopadhyay A. Molecular modeling of the human serotonin1A receptor: role of membrane cholesterol in ligand binding of the receptor. Mol Biosyst. 2011;7:224–34.

29. Sengupta D, Chattopadhyay A. Identification of cholesterol binding sites in the serotonin1A receptor. J Phys Chem B. 2012;116:12991–6.
30. Cherezov V, Rosenbaum DM, Hanson MA, Rasmussen SG, Thian FS, Kobilka TS, Choi HJ, Kuhn P, Weis WI, Kobilka BK, Stevens RC. High resolution crystal structure of an engineered human beta 2 adrenergic G-protein coupled receptoer. Science. 2007;318:1258–65.
31. Liu W, et al. Structural basis for allosteric regulation of GPCRs by sodium ions. Science. 2012;337:232–6.
32. Batyuk A, Galli L, Ishchenko A, Han GW, Gati C, Popov PA, Lee MY, Stauch B, White TA, Barty A, Aquila A, Hunter MS, Lian M, Boutet S, Pu M, Liu ZJ, Nelson G, James D, Li C, Zhao Y, Spence JC, Liu W, Fromme P, Katritch V, Weierstall U, Stevens RC, Cherezov V. Native phasing of x-ray free-electron laser data for a G protein–coupled receptor. Sci Adv. 2016;2:e1600292.
33. Lyman E, et al. A role for a specific cholesterol interaction in stabilizing the Apo configuration of the human A_{2A} adenosine receptor. Structure. 2009;17:1660–8.
34. Lee JYJ, Lyman E. Predictions for cholesterol interaction sites on the A2A adenosine receptor. J Am Chem Soc. 2012;134:16512–5.
35. Rouviere E, Arnarez C, Yang L, Lyman E. Identification of two new cholesterol interaction sites on the A2A adenosine receptor. Biophys J. 2017;113:2415–24.
36. Song Y, Kenwirthy AK, Sanders CR. Cholesterol as a co-solvent and a ligand for membrane proteins. Protein Sci. 2014;23:1–22.
37. Wanling S, Hsin-Yen Y, Robinson CV, Samson MSP. State Dependent lipid interactions with the a2α receptor revealed by md simulations using in vivo–mimetic membranes. doi: https://doi.org/10.1101/362970.
38. Durdagi S, Erol I, Ekhteiari Salmas R, Aksoydan B, Kantarcioglu I. Oligomerization and cooperativity in GPCRs from the perspective of the angiotensin AT1 and dopamine D2 receptors. Neurosci Lett. 2018; https://doi.org/10.1016/j.neulet.2018.04.028.
39. Durdagi S, Aksoydan B, Erol I, Kantarcioglu I, Ergun Y, Bulut G, Acar M, Avsar T, Liapakis G, Karageorgos V, Ekhteiari Salmas R, Sergi B, Alkhatib S, Turan G, Yigit BN, Cantasir K, Kurt B, Kilic T. Integration of multi-scale molecular modeling approaches with experiments for the in silico guided design and discovery of novel herg-neutral antihypertensive oxazalone and imidazolone derivatives and analysis of their potential restrictive effects on cell proliferation. Eur J Med Chem. 2018;145:273–90.
40. Ekhteiari Salmas R, Seeman P, Aksoydan B, Erol I, Kantarcioglu I, Stein M, Yurtsever M, Durdagi S. Analysis of the glutamate agonist LY404,039 binding to non-static dopamine receptor D2 dimer structures and consensus docking. ACS Chem Nerosci. 2017;8:1404–15.
41. Ekhteiari Salmas R, Seeman P, Aksoydan B, Stein M, Yurtsever M, Durdagi S. Biological insights of the dopaminergic stabilizer ACR16 at the binding pocket of dopamine D2 receptor. ACS Chem Nerosci. 2017;8:826–36.
42. Durdagi S, Ekhteiari Salmas R, Stein M, Yurtsever M, Seeman P. Binding interactions of dopamine and apomorphine in d2high and d2low states of human dopamine d2 receptor (d2r) using computational and experimental techniques. ACS Chem Neurosci. 2016;7:185–95.
43. Ayoub MA, Zhang Y, Kelly RS, et al. Functional interaction between angiotensin II receptor type 1 and chemokine (C-C motif) receptor 2 with implications for chronic kidney disease. PLoS One. 2015;10:e0119803.
44. Ferré S. The GPCR heterotetramer: challenging classical pharmacology. Trends Pharmacol Sci. 2015;36:145–52.
45. Terrillon S, Bouvier M. Roles of G-protein-coupled receptor dimerization. EMBO Rep. 2004;5:30–4.
46. Oates J, Watts A. Uncovering the intimate relationship between lipids, cholesterol and GPCR activation. Curr Opin Struct Biol. 2011;21:802–7.
47. Zheng H, Pearsall EA, Hurst DP, Zhang Y, Chu J, Zhou Y, Reggio PH, Loh HH, Law P-Y. Palmitoylation and membrane cholesterol stabilize μ-Opioid receptor homodimerization and G protein coupling. BMC Cell Biol. 2012;13:6.

48. Sengupta D, Kumar GA, Chattopadhyay A. Chapter 16. Interaction of membrane cholesterol with GPCRs: implications in receptor oligomerization. In: Herrick-Davis K, et al., editors. G-protein-coupled receptor dimers, the receptors 33: Springer International Publishing AG; 2017. p. 415. https://doi.org/10.1007/978-3-319-601748_16.

49. Goddard AD, Watts A. Regulation of G protein-coupled receptors by palmitoylation and cholesterol. BMC Biol. 2012;10:27.

50. Periole X. Interplay of G protein-coupled receptors with the membrane: insights from supra-atomic coarse grain molecular dynamics simulations. Chem Rev. 2017;117:156–85.

51. Bukiya AN, Durdagi S, Noskov S, Rosenhouse-Dantsker A. Cholesterol up-regulates neuronal G protein-gated inwardly rectifying potassium (GIRK) channel activity in the hippocampus. J Biol Chem. 2017;292:6135–47.

52. Bolivar JH, Muñoz-García JC, Castro-Dopico T, Dijkman PM, Stansfeld PJ, Watts A. Interaction of lipids with the neurotensin receptor 1. Biochim Biophys Acta. 2016;1858:1278–87.

53. Chakraborty H, Amitabha Chattopadhyay A. Excitements and challenges in GPCR oligomerization: molecular insight from FRET. ACS Chem Nerosci. 2015;6:199–206.

54. Gupta K, Donlan JA, Hopper JT, Uzdavinys P, Landreh M, Struwe WB, Drew D, Baldwin AJ, Stansfeld PJ, Robinson CV. The role of interfacial lipids in stabilizing membrane protein oligomers. Nature. 2017;541:421–4.

55. Gault J, Donlan JA, Liko I, Hopper JT, Gupta K, Housden N, Struwe WB, Marty MT, Mize T, Bechara C, Zhu Y, Wu B, Kleanthous C, Belov M, Damoc E, Makarov A, Robinson CV. High resolution mass spectrometry of small molecules bound to membrane proteins. Nat Methods. 2016;13:333–6.

Cholesterol as a Key Molecule That Regulates TRPV1 Channel Function

Sara L. Morales-Lázaro and Tamara Rosenbaum

Abstract Cholesterol is the one of the major constituents of cell membranes providing these structures with a certain degree of rigidity. Proteins, such as ion channels, are molecules inserted in cell membranes and their activity is regulated by cholesterol and other molecules of a lipidic nature present in them. The molecular mechanisms underlying the regulation of ion channels by lipids and similar molecules have been an object of study for several years. A little over two decades ago, the first mammalian member of the Transient Receptor Potential (TRP) family of ion channels was cloned. This protein, the TRPV1 channel, was shown to integrate several types of noxious signals in sensory neurons and to participate in processes associated to the generation of pain. Thus, TRPV1 has become the target of intense research directed towards finding potential inhibitors of its activity in an effort to control pain. To date, several activators and positive modulators of the activity of TRPV1 have been described. However, very few naturally-occurring inhibitors are known. An endogenously-produced molecule that inhibits the activity of TRPV1 is cholesterol. This chapter focuses on describing the mechanisms by which the activity of TRPV1 can be regulated by this sterol.

Keywords TRPV1 · Cholesterol · Ion channel

1 Introduction

Evolution has allowed different organisms, from invertebrates to vertebrates, to develop the capacity to respond to a wide variety of harmful stimuli for the purpose of preserving their integrity [1]. Ion channels, which are specialized proteins present in the membranes of cells, are among the molecules that allow for the detection of

S. L. Morales-Lázaro (✉) · T. Rosenbaum (✉)
Departamento de Neurociencia Cognitiva, Instituto de Fisiología Celular, Universidad Nacional Autónoma de México, Coyoacan, México City, México
e-mail: saraluzm@ifc.unam.mx; trosenba@ifc.unam.mx

© Springer Nature Switzerland AG 2019
A. Rosenhouse-Dantsker, A. N. Bukiya (eds.), *Direct Mechanisms in Cholesterol Modulation of Protein Function*, Advances in Experimental Medicine and Biology 1135, https://doi.org/10.1007/978-3-030-14265-0_6

such noxious signals. Some of these ion channels can be activated, in a polymodal fashion, by thermal, chemical and mechanical signals. The activation of these multimeric proteins allows the fast passive diffusion of ions across cell membranes, converting different noxious messages into electrical signals [1].

The family of Transient Receptor Potential (TRP) non-selective cation channels are classified into seven subfamilies: TRPA (Ankyrin), TRPC (Canonical), TRPM (Melastatin), TRPML (Mucolipin), TRPP (Polycystic), TRPV (Vanilloid) and TRPN (no mechanoreceptor) [2]. Most of these channels have been described in mammals, however the TRPN channel has only been identified in insects, nematodes, zebrafish and amphibians [2].

TRP channels are structurally similar to voltage-gated ion channels [3], consisting of four subunits that are associated as homo or heterotretramers [4] (Fig. 1a). Each subunit is a protein that consists of six transmembrane domains (S1-6), with amino- and carboxyl-termini located intracellularly and an ionic conduction pore formed in the tetramer by the linker located between the S5 and S6 [4] (Fig. 1b). As in all other proteins, TRP channels exhibit a tight relationship between their structure and their function and contain amino acid residues that give rise to regions that constitute activation sites for the channels by diverse stimuli.

TRP channels of mammals are abundantly expressed in sensory neurons from dorsal root and trigeminal ganglia (DRG and TG, respectively), among other several types of cells, where they enable the detection of harmful signals [5]. Among these TRP channels, we find TRPA1, TRPM8, TRPM3 and some members of the vanilloid receptors (TRPV1-4) [5, 6]. All of these proteins are termed thermo-TRP channels because they can be activated by cold (TRPA1 and TRPM8) [7, 8] or by warm or hot temperatures (TRPM3, TRPV1-TRPV4) [6, 9–12]. These polymodal ion channels can also be activated by several chemical compounds found in plants. For example, TRPA1 is activated by isothiocyanates in garlic and mustard oil [13]; TRPM8 is acti-

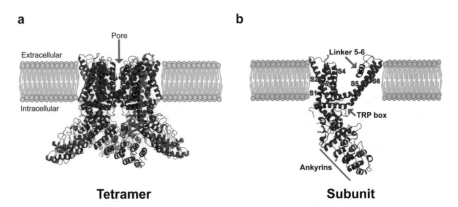

Fig. 1 TRPV1 ion channel topology. (**a**) Fourfold symmetry of TRP ion channel surrounded a pore the ion pore conduction (modified from PDB 3J5P). (**b**) The TRP subunit is composed of six transmembrane segments (S1–S6), an amino with a segment of six ankyrin repeats and carboxyl-end located intracellularly. The external pore the ion-conductivity pore is lined by linker between S5–6

vated by menthol [8] and TRPV1 is activated by capsaicin, the pungent compounds found in chili peppers of the *Capsicum* genus [10]. Additionally, some endogenous compounds (of a lipidic and/or peptidic nature) released during tissue injury or inflammation, are potent activators of these channels [14]. Thus, pathological states where there is an upregulation of these inflammatory mediators are accompanied by pain due to the activation of these ion channels located in the surface of sensory neurons.

To date, a large number of compounds that activate these channels has been identified and there is ongoing research focused on finding inhibitory molecules, which will be of therapeutic value if they can counteract the pain associated with the activation of these proteins [15]. TRPV1 has been the most studied member of the TRP-channel family due to its roles in acute and chronic pain, as will be discussed in detail in the next section.

2 TRPV1: More than a Capsaicin Receptor

The vanilloid subfamily contains the better studied member of the family of TRP ion channels, the capsaicin receptor (TRPV1). This receptor was first cloned in 1997 [10] and identified as a channel directly activated or modulated by several noxious inputs such as hot temperatures (≥ 42 °C), natural irritant compounds (i.e., capsaicin, allicin), toxins from plants or spider venoms (resiniferatoxin and double-Knot toxin, respectively), extracellular acid or intracellular alkaline pH and by endogenous lipid mediators such as anandamide, lysophosphatidic acid (LPA), diacylglycerol, phosphatidylinositol 4,5-biphosphate (PIP$_2$), and arachidonic acid or lipoxygenase products [10, 16].

Furthermore, the first high-resolution structure of TRP channels to be resolved was that of TRPV1 [17]. The resolved three-dimensional structure was obtained by using single particle electron cryomicroscopy (cryo-EM), corroborating that TRPV1 exhibits a fourfold symmetry that surrounds a central pore formed by the extracellular linker located between S5–S6 [17] (Fig. 1a). The S1–S4 voltage-sensing like domains are contiguous to S5-S6 and tetrameric organization is enabled by interactions between the intracellular domains: the amino and carboxy ends [17]. Interestingly, S1–S4 domains provide a long surface for the interaction of lipophilic ligands such as capsaicin, anandamide, resiniferatoxin or lipid mediators [17] and allicin binds to an ankyrin repeat domain in the N-terminus [18].

In this regard, our group has described that oleic acid, a lipophilic ligand, binds to the same pocket as capsaicin does; however, in contrast to capsaicin, the effect of this interaction is the inhibition of the activity of TRPV1, constituting one of the scarce antagonists described for this channel [19]. In this respect, it has also been reported that another lipophilic molecule, namely cholesterol, modulates TRPV1 function and such modulation has been proposed as a two-pronged mechanism: cholesterol indirectly regulates TRPV1 through changes in the properties of the membrane [20–23] or through a direct interaction with a region in the structure of TRPV1 [24, 25]. The following sections will describe the current knowledge on the effects of cholesterol on TRPV1 function.

3 Effects of Membrane Cholesterol Depletion on TRPV1 Function

Plasma membranes are composed of lipid molecules and proteins and the major lipid constituent in animal membranes is cholesterol, an amphipathic molecule formed by a rigid planar tetracyclic ring with an angular methyl group on a side, an aliphatic chain (isooctyl) attached to C7 and a β-hydroxyl group at the C3 position [26]. The latter is a functional group important for H-bond formation.

Cholesterol distribution in the plasma membrane is confined to microdomains called lipid rafts, which are cholesterol- and sphingolipid-enriched membrane domains characterized by their insolubility in Triton X-100 detergent, thus, they are also known as detergent-resistant membranes (DRMs) [27]. These membrane rafts form functional platforms for regulating signal pathways where some protein receptors are clustered [28]. Interestingly, some ligand-gated receptors, such as some ion channels, have been shown to be localized in these kinds of microdomains [29]. For example, the TRPV1 ion channel has been suggested to be localized in lipid rafts from DRG [21] or TG neurons, while transiently-transfected TRPV1 in cell lines like HEK293, has not been shown to be present in these specialized membrane domains [24].

Experimental evidences show that membrane depletion of cholesterol modifies the activation of TRPV1 by capsaicin and protons [21]. This was determined by performing whole-cell patch-clamp recordings in DRG neurons pretreated with methyl β-cyclodextrin (MβCD, which removes membrane-bound cholesterol). In these experiments, this pretreatment significantly decreased the amplitude of capsaicin- and proton-evoked currents as compared to untreated neurons [21]. Additionally, TRPV1 clusters localized to the surface of DRG neurons was drastically decreased when neurons were depleted of cholesterol [21]. Thus, the integrity of cholesterol in the plasma membranes of cells determines the appropriate localization and function of the TRPV1 channel.

Similarly, it was demonstrated that the effects of capsaicin and resiniferatoxin are abrogated in neurons from trigeminal ganglia depleted of cholesterol, since Ca^{2+} influx decreased in cells pretreated with MβCD [22]. Notably, this decrease was also observed when the neurons were treated with compounds that disrupt other lipid raft components such as sphingomyelin, indicating that lipid raft disruption affects Ca^{2+} influx related to the activation of TRPV1 [22].

In addition to these observations, recently it has been shown that rat TRPV1 channels transiently expressed in F11 (rat embryonic dorsal root ganglion) cells, co-localized with specific markers of lipid rafts such as flotillin [25], confirming the presence of TRPV1 in these specialized microdomains. Thus, lipid raft disruption could be a key way for the attenuation of TRPV1 activation and of the physiological effects linked to the function of this ion channel.

These experimental evidences show that TRPV1 localization to DMRs is dependent upon the presence of cholesterol in the membranes from DRG neurons, the native expression system. However, under experimental conditions where there is transient expression of TRPV1 in epithelial cell lines, such as HEK293 cells, TRPV1 has not been found in these specialized microdomains [24]. These could be explained by the fact that different cells exhibit varying cholesterol contents. It is possible that neurons are a subtype of cells that contain higher cholesterol levels in their plasma membranes as compared to other types of cells and that cholesterol depletion leads to severe changes on the localization of particular proteins.

4   The TRPV1 Channel is Regulated by Cholesterol Via Specific Interactions

To date, numerous ion channels whose function can be modified by cholesterol have been identified. However, there are few channels have been shown to be regulated by this sterol through direct and specific interactions [30]. Among these channels are some potassium (K_{ir}, BK_{Ca}) channels [31–34], ligand-gated channels (GABA, nAChR, P2X) [35–37], and TRPV1 [24].

Specific interactions of cholesterol with ion channels are largely attributed to the presence of CRACs, which stands for Cholesterol Recognition/interaction Amino acid Consensus []. This short linear motif has the sequence (L/V)-X_{1-5}-(Y)-X_{1-5}-(K/R), where X represents any of one to five amino acids [38]. In addition, cholesterol can bind to an ion channel by interacting with an inverted CRAC motif named the CARC domain. This motif (K/R)-X_{1-5}-(Y/F)-X_{1-5}-(L/V) [39] is different from the CRAC sequence in its orientation and because the central amino acid can be either a tyrosine (Y) or a phenylalanine (F) [39]. Moreover, another cholesterol binding motif has been described as the Cholesterol Consensus Motif (CCM) that, unlike the other two motifs whose sequences are found in a continuous segment, the sequence of the CCM motif is bipartite, being located in two adjacent helices: the first sequence (W/Y)-(I/V/L)-(K/R) found on one helix and the amino acid (F/Y/R) on the other helix [30, 40].

Some of these cholesterol-binding motifs have only been described for two members of the vanilloid receptors: TRPV1 and TRPV4 [24, 25, 41]; however, a direct interaction of the lipid and a TRP channel, as well as the consequences on the activation of the protein, has only been demonstrated experimentally for TRPV1 [24].

Previous reports using whole-cell experiments in DRG neurons depleted of cholesterol yielded two possibilities to explain the reduction on capsaicin generated current densities: alteration of the activity of TRPV1 in the plasma membrane or modifications on channel trafficking to the membrane. In a study performed by our research group, we sought to examine the effects of cholesterol on TRPV1 channels using excised membranes from HEK293 cells expressing this channel. These experiments provided a tool to exclude the possible involvement of cholesterol on cell-trafficking mechanisms and to explore direct effects of the sterol on the ion channels. Depletion of cholesterol from these membrane patches with MβCD showed that TRPV1 activation by capsaicin was not affected by this treatment, as assessed using inside-out patches with the patch clamp technique. However, the addition of cholesterol to inside-out membrane patches produced a significant reduction in rat TRPV1 activation by capsaicin, temperature and voltage. Since cholesterol enrichment did not modify parameters such as the Hill coefficient and the $K_{1/2}$ for activation by capsaicin, we concluded that binding of this agonist to TRPV1 was unmodified in the presence of cholesterol and not responsible for the effects of this lipid on channel function [24] .

To further detail the effects of cholesterol on rat TRPV1 function, we performed noise analysis experiments in inside-out patches from HEK293 cells expressing TRPV1 before and after cholesterol depletion in these excised patches. The results showed that the open probability (Po) of TRPV1 remained similar before and after cholesterol enrichment. However, the number of functional channels (N) decreased considerably [24]. These results were confirmed by excised patch single channel recordings, where cholesterol did not alter the magnitude of the single-channel currents nor the open probability of the channel. Nonetheless, after a few minutes of membrane enrichment with cholesterol we observed that the number of capsaicin-responsive channels decreased during our electrophysiological recordings [24].

In order to determine whether the reduction of the number of functional TRPV1 channels in cholesterol-enriched membranes was due to specific interactions of cholesterol with the channel, we searched sequences that resembled a possible CRAC motif and we found an inverted CRAC sequence (CARC motif) located in the S5 of TRPV1: $\mathbf{R_{579}}F_{580}M_{581}\mathbf{F_{582}}V_{583}Y_{584}\mathbf{L_{585}}$ [24] (Fig. 2). We next evaluated the effects of cholesterol on rat TRPV1 mutant versions where the amino acids corresponding to the CARC motif were individually substituted by other residues. Our experiments showed that a charge reversal of the positive amino acid R576 to an aspartate (D), and the change of the aromatic F582 to a polar amino acid such as glutamine (Q), partially inhibited the effects of cholesterol on TRPV1 function [24]. Moreover, when L585 was changed to an isoleucine (I), the inhibitory effects of cholesterol on TRPV1 activity were completely abolished.

In order to demonstrate the stereospecificity of the interaction between cholesterol and TRPV1, we used a synthetic chiral cholesterol analogue, epicholesterol, which has the hydroxyl group in position 3α instead of position 3β. Membrane patch

Fig. 2 Alignment sequence between the CRAC and CARC domains from rat and human. Consensus sequence for the CARC motif found in rat TRPV1 channel, (NP_114188) alignment with the from human TRPV1 isoform 1 (iso1-hTRPV1) (NP_542435) and the genetic variant from human TRPV1, var-hTRPV1 (ABA06605.1). The alignment also shows that the CRAC motif is conserved between human and rat TRPV1 channels

incubation with epicholesterol did not modify capsaicin evoked currents, indicating that the effects of cholesterol on rat TRPV1 function are stereospecific [24].

The specificity of cholesterol interactions within the TRPV1 channel was simulated by molecular docking using the recently reported TRPV1 cryo-EM structure (Fig. 3). These molecular simulations suggest that TRPV1 interactions with cholesterol are mediated by CH-π stacking between the rings of cholesterol (α-face) and the aromatic ring of the tyrosine or phenylalanine residues located in the middle of the CARC domain in the channel; whereas leucine or valine make Van der Waals contacts with the aliphatic chains (methyl and isooctyl groups) located in the β-face of cholesterol (Fig. 3). Although the β-OH group of cholesterol on the CRAC motif is putatively capable of forming H-bonds with the phenol group of the tyrosine, this is impossible for the CARC motif (contained in the TRPV1 channel), since the aromatic ring of the phenylalanine lacks a hydroxyl group. However, it is possible that the hydroxyl group of cholesterol forms a H-bond with R585 at the inner leaflet, as has been recently described for other arginine residues located at the inner leaflet within the TRPV1 channel [25].

Finally, we also evaluated cholesterol effects on the human TRPV1 channel and found that the isoform 1 of human TRPV1 (iso1-hTRPV1, NP_542435) lacks the leucine in position 585 that is present in the rat TRPV1 sequence, Fig. 2. By further examining the literature on reported human TRPV1 sequences, we found a genetic variant (var-hTRPV1) with a single nucleotide polymorphism that contained a valine in this position (SNP: rs8065080) [42] (Fig. 2). This is an amino acid that actually forms part of the CARC motif, so we hypothesized that this variant would exhibit a similar behavior in response to cholesterol to the one we had observed in the rat TRPV1. Furthermore, isoform 1 human TRPV1 contains an isoleucine at position 585, a residue similar to the one we had introduced in the rat TRPV1 (L585I) and that

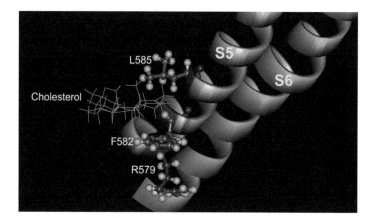

Fig. 3 Molecular docking simulation of cholesterol bound to TRPV1. Simulation using the cryo-EM structure of TRPV1 in the closed state (PDB 3J5P, chain a). Docking simulation shows the proximity between cholesterol and amino acids F582 and L585 located in the CARC motif of TRPV1

had produced channels resistant to inhibition by cholesterol. Indeed, it was not surprising to find that activation of the human TRPV1 isoform 1 was not inhibited by cholesterol. We also found that the substitution in this isoform 1 of I585 by a valine (as the one present in var-hTRPV1) rendered the channel partially susceptible to cholesterol [24]. Furthermore, when I585 from the iso1-hTRPV1 was mutated to leucine (emulating the amino acid contained in the rat TRPV1), this amino acid substitution conferred a marked susceptibility to cholesterol, producing iso1-hTRPV1-I585L channels that were inhibited in their activation by capsaicin when the membrane patches were enriched with this sterol [24].

The iso1-hTRPV1 does not contain a CARC domain at the same position where the rat TRPV1 CARC domain is located. So, it is possible that human or rat TRPV1 channels contain CRAC, CARC or CCM motifs in other sequence positions. Consistent with this idea, it was recently reported that human TRPV1 has two CRAC motifs (located in aa's 349-356 and 553-557), two CARC motifs (located in aa's 304-316 and 535-542) and one CCM motif (in aa's 433-447). Of these cholesterol-binding regions, the CRAC motif located in the S4 and linker S4-S5 region (553-557 aa) (Fig. 2) and the CCM motif, are the most conserved between TRPV1 channels from several species [25]. However, until now, there are only docking simulations showing the interaction of cholesterol with R557 located in the conserved CRAC motif and the functional significances of these motifs remain unresolved [25].

5   Cholesterol Depletion Modifies TRPV1 Permeability to Large-Cations

Plasma membranes properties can be modified according to their lipid composition (i.e. by cholesterol depletion or enrichment) [43]. Two important features of ion channels are their gating properties and their ion selectivity and both are highly dependent upon the membrane lipid content.

Ion selectivity has been considered as a fixed parameter for each channel, however, it has been described that some of them have dynamic pores, leading permeability to large cations, this process has been known as pore dilation [44–46]. Although, this concept has been reconsidered [47], there are experimental data demonstrating that some ion channels, such as TRPV1, can permeate large molecules under specific conditions.

In 2003 Meyers et al., provided the first evidence for the permeation of a large molecule, the nontoxic fluorescent cationic dye (FM1-43,) through the pore of TRPV1 [48]. Furthermore, Binshtok et al., demonstrated that when TRPV1 was activated by capsaicin, the permeation of a local cationic anesthetic QX-314 (a Na^+ ion channel blocker) is allowed, producing local pain inhibition without affecting other sensations [49]. Additionally, Chung et al., found that sustained TRPV1 exposure to capsaicin or protons induced a dilated pore allowing for the uptake of YO-PRO1 and FM1-43 into cells [45]. All these examples show how the TRPV1 can be modified, rendering the channel permeable to the passage of large cationic molecules.

Interestingly, these dynamic changes in the TRPV1 pore are susceptible to changes in the cholesterol content of membranes [20]. This was demonstrated in CHO cells (Chinese hamster ovary cells) with inducible human TRPV1 expression. The cholesterol content in these cells was depleted by ~54% through the treatment with MβCD:cholesterol (10:1) and whole cell patch clamp recordings were used to evaluate the effects of cholesterol depletion on the permeability to N-methyl-D-Glucamine (NMDG, a large monovalent cation) [20]. Sustained TRPV1 activation by capsaicin, under hypocalcemic conditions in cells depleted of cholesterol, decreased permeability to NMDG, indicating that cholesterol reduction in membrane affects the ion-permeability properties of the channel [20]. In addition, it has been shown that sustained TRPV1 activation by capsaicin in CHO cells, leads to YO-PRO1 uptake and cholesterol depletion decreases the uptake of this cationic large dye [20]. Similar effects were observed whether the channels are activated by protons under sustained and hypocalcemic conditions [20]. In contrast to these results, the permeability to these large cationic molecules in cells depleted of cholesterol were unaffected if TRPV1 was activated by temperature [20], indicating that the properties on TRPV1 ion-permeability also depend on the type of noxious stimulus that activates this channel.

6 Cholesterol Effects on TRPV1 Temperature Responses

We have discussed that cholesterol has strong influences on membrane fluidity [50] which affect the localization or function of transmembrane proteins as ion channels [43] but the amount of cholesterol in membranes also modifies the sensitivity of certain receptors to specific stimuli. For example, TRPV3 is also a thermosensitive channel [51] and it was reported that TRPV3 channels are sensitized to activation by lower temperatures (below 30 °C) in cholesterol-enriched cells [52].

Effects of cholesterol on TRPV1's temperature-sensitivity have been also evaluated [23]. Whole-cell recordings from TRPV1-expressing HEK293 cells with cholesterol enhancement showed that the temperature threshold for TRPV1 activation was significantly increased, since the half-activation temperature was 50 °C in comparison to 48 °C in untreated cells. Notably, the heat response of TRPV1 was unchanged in cholesterol-depleted cells. Therefore, the temperature threshold for TRPV1 activation is partially modified only when this sterol is increased in the membrane.

Thus, it is interesting to note that two structurally-related ion channels, TRPV3 and TRPV1, show distinct responses to cholesterol concentrations in the plasma membrane. Although it is not clear why these channels exhibit different responses to cholesterol, it is not uncommon that, even closely-related ion channels, respond differently to the same molecule. For example, the energetic transitions coupled to the opening or closing of the pore of a specific ion channel are not necessarily altered in the same way as what occurs in another ion channel. Moreover, in the particular cases of TRPV1 and TRPV3, the mechanism by which cholesterol modulates thermal sensitivity of TRPV3 is unknown and could very well be through effects on membrane rigidity and not through a direct a mechanism as the one described for TRPV1.

7 Conclusion

Fifty years ago, it was unimaginable to consider cholesterol as a regulator of ion channels, since this sterol was only considered as a key component of the lipid bilayer where it maintains direct interaction with phospholipids. There is no doubt that cholesterol content modifies the properties of the membranes, having a strong influence in the regulation on ion channel function.

Furthermore, this sterol can regulate the function of these integral membrane proteins in a direct and specific fashion: through the interaction with specific amino acids within the sequence of these proteins.

This chapter has described the molecular mechanisms underlying the regulation of TRPV1 function by changes of cholesterol content in the cell plasma membrane or by direct interaction between the channel and the sterol.

Since cholesterol depletion modifies TRPV1 localization in the plasma membranes from sensory neurons affecting the calcium influx [21, 22], it has been concluded that TRPV1 is confined to specialized microdomains of the membrane: lipid rafts. The discrepancies observed in some cell lines where TRPV1 channels has not been identified in these detergent resistant domains [22, 24] could be attributed to differences in the amount of cholesterol between cell types. Moreover, cholesterol depletion from cells also modulates TRPV1 ion-permeability [20], indicating that the enlargement of the TRPV1 pore is strongly dependent upon cholesterol content. This differs from the idea that cholesterol only promotes stiffness of cellular membranes, since for the case of TRPV1, cholesterol has an important role on the elasticity of its pore. In addition, we have also discussed that cholesterol enrichment causes changes in TRPV1's threshold to heat [23]. Together, these experimental demonstrations, show that TRPV1 is highly susceptible to be regulated by changes on the amount of cholesterol in the membranes where it is expressed.

We have also detailed on how cholesterol directly inhibits TRPV1 activation by capsaicin, temperature and voltage through the specific interaction of cholesterol in a CARC motif located on the S4 from TRPV1 [24] and on the different TRPV1 susceptibilities to cholesterol of different species (rat and human isoform or variants) [36]. Data from our work group and further examination of the literature, has led us to conclude that cholesterol can inhibit human TRPV1 activation by interacting with other cholesterol binding domains recently described [25].

Now, cholesterol is considered as a molecule key in regulating TRPV1 function; however, the physiological consequences of this regulation have still to be explored.

Acknowledgments This work was supported by grants from Dirección General de Asuntos del Personal Académico (DGAPA)-Programa de Apoyo a Proyectos de Investigación e Innovación Tecnológica (PAPIIT) IN206819 and by Estímulos a Investigaciones Médicas Miguel Alemán Valdés to S.LM.L. and Consejo Nacional de Ciencia y Tecnología (CONACyT) A1-S-8760, grant from Fronteras en la Ciencia No. 77 from CONACyT and DGAPA-PAPIIT IN200717 to T.R.

References

1. Smith ES, Lewin GR. Nociceptors: a phylogenetic view. J Comp Physiol A Neuroethol Sens Neural Behav Physiol. 2009;195:1089–106. https://doi.org/10.1007/s00359-009-0482-z.
2. Li H. TRP channel classification. Adv Exp Med Biol. 2017;976:1–8. https://doi.org/10.1007/978-94-024-1088-4_1.
3. Jiang Y, Lee A, Chen J, Ruta V, Cadene M, Chait BT, et al. X-ray structure of a voltage-dependent K+ channel. Nature. 2003;423:33–41. https://doi.org/10.1038/nature01580.
4. Ramsey IS, Delling M, Clapham DE. An introduction to TRP channels. Annu Rev Physiol. 2006;68:619–47. https://doi.org/10.1146/annurev.physiol.68.040204.100431.
5. Mickle AD, Shepherd AJ, Mohapatra DP. Nociceptive TRP channels: sensory detectors and transducers in multiple pain pathologies. Pharmaceuticals (Basel). 2016;9:E72. https://doi.org/10.3390/ph9040072.
6. Vandewauw I, De Clercq K, Mulier M, Held K, Pinto S, Van Ranst N, et al. A TRP channel trio mediates acute noxious heat sensing. Nature. 2018;555:662–6. https://doi.org/10.1038/nature26137.
7. Story GM, Peier AM, Reeve AJ, Eid SR, Mosbacher J, Hricik TR, et al. ANKTM1, a TRP-like channel expressed in nociceptive neurons, is activated by cold temperatures. Cell. 2003;112:819–29.
8. Bautista DM, Siemens J, Glazer JM, Tsuruda PR, Basbaum AI, Stucky CL, et al. The menthol receptor TRPM8 is the principal detector of environmental cold. Nature. 2007;448:204–8. https://doi.org/10.1038/nature05910.
9. Smith GD, Gunthorpe MJ, Kelsell RE, Hayes PD, Reilly P, Facer P, et al. TRPV3 is a temperature-sensitive vanilloid receptor-like protein. Nature. 2002;418:186–90. https://doi.org/10.1038/nature00894.
10. Caterina MJ, Schumacher MA, Tominaga M, Rosen TA, Levine JD, Julius D. The capsaicin receptor: a heat-activated ion channel in the pain pathway. Nature. 1997;389:816–24. https://doi.org/10.1038/39807.
11. Guler AD, Lee H, Iida T, Shimizu I, Tominaga M, Caterina M. Heat-evoked activation of the ion channel, TRPV4. J Neurosci. 2002;22:6408–14.
12. Caterina MJ, Rosen TA, Tominaga M, Brake AJ, Julius D. A capsaicin-receptor homologue with a high threshold for noxious heat. Nature. 1999;398:436–41. https://doi.org/10.1038/18906.
13. Jordt SE, Bautista DM, Chuang HH, McKemy DD, Zygmunt PM, Hogestatt ED, et al. Mustard oils and cannabinoids excite sensory nerve fibres through the TRP channel ANKTM1. Nature. 2004;427:260–5. https://doi.org/10.1038/nature02282.
14. Taberner FJ, Fernandez-Ballester G, Fernandez-Carvajal A, Ferrer-Montiel A. TRP channels interaction with lipids and its implications in disease. Biochim Biophys Acta. 2015;1848:1818–27. https://doi.org/10.1016/j.bbamem.2015.03.022.
15. Levine JD, Alessandri-Haber N. TRP channels: targets for the relief of pain. Biochim Biophys Acta. 2007;1772:989–1003. https://doi.org/10.1016/j.bbadis.2007.01.008.
16. Morales-Lazaro SL, Simon SA, Rosenbaum T. The role of endogenous molecules in modulating pain through transient receptor potential vanilloid 1 (TRPV1). J Physiol. 2013;591:3109–21. https://doi.org/10.1113/jphysiol.2013.251751.

17. Liao M, Cao E, Julius D, Cheng Y. Structure of the TRPV1 ion channel determined by electron cryo-microscopy. Nature. 2013;504:107–12. https://doi.org/10.1038/nature12822.
18. Salazar H, Llorente I, Jara-Oseguera A, Garcia-Villegas R, Munari M, Gordon SE, et al. A single N-terminal cysteine in TRPV1 determines activation by pungent compounds from onion and garlic. Nat Neurosci. 2008;11:255–61. https://doi.org/10.1038/nn2056.
19. Morales-Lazaro SL, Llorente I, Sierra-Ramirez F, Lopez-Romero AE, Ortiz-Renteria M, Serrano-Flores B, et al. Inhibition of TRPV1 channels by a naturally occurring omega-9 fatty acid reduces pain and itch. Nat Commun. 2016;7:13092. https://doi.org/10.1038/ncomms13092.
20. Jansson ET, Trkulja CL, Ahemaiti A, Millingen M, Jeffries GD, Jardemark K, et al. Effect of cholesterol depletion on the pore dilation of TRPV1. Mol Pain. 2013;9:1. https://doi.org/10.1186/1744-8069-9-1.
21. Liu M, Huang W, Wu D, Priestley JV. TRPV1, but not P2X, requires cholesterol for its function and membrane expression in rat nociceptors. Eur J Neurosci. 2006;24:1–6. https://doi.org/10.1111/j.1460-9568.2006.04889.x.
22. Szoke E, Borzsei R, Toth DM, Lengl O, Helyes Z, Sandor Z, et al. Effect of lipid raft disruption on TRPV1 receptor activation of trigeminal sensory neurons and transfected cell line. Eur J Pharmacol. 2010;628:67–74. https://doi.org/10.1016/j.ejphar.2009.11.052.
23. Liu B, Hui K, Qin F. Thermodynamics of heat activation of single capsaicin ion channels VR1. Biophys J. 2003;85:2988–3006. https://doi.org/10.1016/S0006-3495(03)74719-5.
24. Picazo-Juarez G, Romero-Suarez S, Nieto-Posadas A, Llorente I, Jara-Oseguera A, Briggs M, et al. Identification of a binding motif in the S5 helix that confers cholesterol sensitivity to the TRPV1 ion channel. J Biol Chem. 2011;286:24966–76. https://doi.org/10.1074/jbc.M111.237537.
25. Saha S, Ghosh A, Tiwari N, Kumar A, Kumar A, Goswami C. Preferential selection of Arginine at the lipid-water-interface of TRPV1 during vertebrate evolution correlates with its snorkeling behaviour and cholesterol interaction. Sci Rep. 2017;7:16,808. https://doi.org/10.1038/s41598-017-16780-w.
26. Song Y, Kenworthy AK, Sanders CR. Cholesterol as a co-solvent and a ligand for membrane proteins. Protein Sci. 2014;23:1–22. Epub 2013/10/25. https://doi.org/10.1002/pro.2385.
27. Rietveld A, Simons K. The differential miscibility of lipids as the basis for the formation of functional membrane rafts. Biochim Biophys Acta. 1998;1376:467–79.. Epub 1998/11/07
28. Lingwood D, Simons K. Lipid rafts as a membrane-organizing principle. Science. 2010;327:46–50. Epub 2010/01/02. https://doi.org/10.1126/science.1174621.
29. Dart C. Lipid microdomains and the regulation of ion channel function. J Physiol. 2010;588:3169–78. Epub 2010/06/04. https://doi.org/10.1113/jphysiol.2010.191585.
30. Levitan I, Singh DK, Rosenhouse-Dantsker A. Cholesterol binding to ion channels. Front Physiol. 2014;5:65. https://doi.org/10.3389/fphys.2014.00065.
31. Romanenko VG, Rothblat GH, Levitan I. Modulation of endothelial inward-rectifier K+ current by optical isomers of cholesterol. Biophys J. 2002;83:3211–22. https://doi.org/10.1016/S0006-3495(02)75323-X.
32. Singh DK, Rosenhouse-Dantsker A, Nichols CG, Enkvetchakul D, Levitan I. Direct regulation of prokaryotic Kir channel by cholesterol. J Biol Chem. 2009;284:30,727–36. https://doi.org/10.1074/jbc.M109.011221.
33. Rosenhouse-Dantsker A, Noskov S, Durdagi S, Logothetis DE, Levitan I. Identification of novel cholesterol-binding regions in Kir2 channels. J Biol Chem. 2013;288:31,154–64. https://doi.org/10.1074/jbc.M113.496117.
34. Bukiya AN, Belani JD, Rychnovsky S, Dopico AM. Specificity of cholesterol and analogs to modulate BK channels points to direct sterol-channel protein interactions. J Gen Physiol. 2011;137:93–110. https://doi.org/10.1085/jgp.201010519.
35. Addona GH, Sandermann H Jr, Kloczewiak MA, Miller KW. Low chemical specificity of the nicotinic acetylcholine receptor sterol activation site. Biochim Biophys Acta. 2003;1609:177–82.

36. Sooksawate T, Simmonds MA. Effects of membrane cholesterol on the sensitivity of the GABA(A) receptor to GABA in acutely dissociated rat hippocampal neurones. Neuropharmacology. 2001;40:178–84.
37. Robinson LE, Shridar M, Smith P, Murrell-Lagnado RD. Plasma membrane cholesterol as a regulator of human and rodent P2X7 receptor activation and sensitization. J Biol Chem. 2014;289:31983–94. https://doi.org/10.1074/jbc.M114.574699.
38. Li H, Papadopoulos V. Peripheral-type benzodiazepine receptor function in cholesterol transport. Identification of a putative cholesterol recognition/interaction amino acid sequence and consensus pattern. Endocrinology. 1998;139:4991–7. https://doi.org/10.1210/endo.139.12.6390.
39. Baier CJ, Fantini J, Barrantes FJ. Disclosure of cholesterol recognition motifs in transmembrane domains of the human nicotinic acetylcholine receptor. Sci Rep. 2011;1:69. https://doi.org/10.1038/srep00069.
40. Hanson MA, Cherezov V, Griffith MT, Roth CB, Jaakola VP, Chien EY, et al. A specific cholesterol binding site is established by the 2.8 A structure of the human beta2-adrenergic receptor. Structure. 2008;16:897–905. https://doi.org/10.1016/j.str.2008.05.001.
41. Kumari S, Kumar A, Sardar P, Yadav M, Majhi RK, Kumar A, et al. Influence of membrane cholesterol in the molecular evolution and functional regulation of TRPV4. Biochem Biophys Res Commun. 2015;456:312–9. https://doi.org/10.1016/j.bbrc.2014.11.077.
42. Cantero-Recasens G, Gonzalez JR, Fandos C, Duran-Tauleria E, Smit LA, Kauffmann F, et al. Loss of function of transient receptor potential vanilloid 1 (TRPV1) genetic variant is associated with lower risk of active childhood asthma. J Biol Chem. 2010;285:27532–5. https://doi.org/10.1074/jbc.C110.159491.
43. McIntosh TJ, Simon SA. Roles of bilayer material properties in function and distribution of membrane proteins. Annu Rev Biophys Biomol Struct. 2006;35:177–98. https://doi.org/10.1146/annurev.biophys.35.040405.102022.
44. Virginio C, MacKenzie A, Rassendren FA, North RA, Surprenant A. Pore dilation of neuronal P2X receptor channels. Nat Neurosci. 1999;2:315–21. https://doi.org/10.1038/7225.
45. Chung MK, Guler AD, Caterina MJ. TRPV1 shows dynamic ionic selectivity during agonist stimulation. Nat Neurosci. 2008;11:555–64. https://doi.org/10.1038/nn.2102.
46. Chen J, Kim D, Bianchi BR, Cavanaugh EJ, Faltynek CR, Kym PR, et al. Pore dilation occurs in TRPA1 but not in TRPM8 channels. Mol Pain. 2009;5:3. https://doi.org/10.1186/1744-8069-5-3.
47. Li M, Toombes GE, Silberberg SD, Swartz KJ. Physical basis of apparent pore dilation of ATP-activated P2X receptor channels. Nat Neurosci. 2015;18:1577–83. https://doi.org/10.1038/nn.4120.
48. Meyers JR, MacDonald RB, Duggan A, Lenzi D, Standaert DG, Corwin JT, et al. Lighting up the senses: FM1-43 loading of sensory cells through nonselective ion channels. J Neurosci. 2003;23:4054–65.
49. Binshtok AM, Bean BP, Woolf CJ. Inhibition of nociceptors by TRPV1-mediated entry of impermeant sodium channel blockers. Nature. 2007;449:607–10. https://doi.org/10.1038/nature06191.
50. de Meyer F, Smit B. Effect of cholesterol on the structure of a phospholipid bilayer. Proc Natl Acad Sci U S A. 2009;106:3654–8. https://doi.org/10.1073/pnas.0809959106.
51. Liu B, Qin F. Single-residue molecular switch for high-temperature dependence of vanilloid receptor TRPV3. Proc Natl Acad Sci U S A. 2017;114:1589–94. https://doi.org/10.1073/pnas.1615304114.
52. Klein AS, Tannert A, Schaefer M. Cholesterol sensitises the transient receptor potential channel TRPV3 to lower temperatures and activator concentrations. Cell Calcium. 2014;55:59–68. https://doi.org/10.1016/j.ceca.2013.12.001.

Cholesterol Binding Sites in Inwardly Rectifying Potassium Channels

Avia Rosenhouse-Dantsker

Abstract Inwardly rectifying potassium (Kir) channels play a variety of critical cellular roles including modulating membrane excitability in neurons, cardiomyocytes and muscle cells, and setting the resting membrane potential, heart rate, vascular tone, insulin release, and salt flow across epithelia. These processes are regulated by a variegated list of modulators. In particular, in recent years, cholesterol has been shown to modulate a growing number of Kir channels. Subsequent to the discovery that members of the Kir2 subfamily were down-regulated by cholesterol, we have shown that members of several other Kir subfamilies were also modulated by cholesterol. However, not all cholesterol sensitive Kir channels were down-regulated by cholesterol. Our recent studies focused on three Kir channels: Kir2.1 (IRK1), Kir3.2^ (GIRK2^) and Kir3.4* (GIRK4*). Among these, Kir2.1 was down-regulated by cholesterol whereas Kir3.2^ and Kir3.4* were both up-regulated by cholesterol. Despite the opposite impact of cholesterol on these Kir3 channels compared to Kir2.1, putative cholesterol binding sites in all three channels were identified in equivalent transmembrane domains. Interestingly, however, there are intriguing differences in the specific residues that interact with the cholesterol molecule in these Kir channels. Here we compare and contrast the molecular characteristics of the putative cholesterol binding sites in the three channels, and discuss the potential implications of the differences for the impact of cholesterol on ion channels.

Keywords Cholesterol · Kir channels · GIRK channels · Ion channels · Channel modulation · Protein–lipid interaction

A. Rosenhouse-Dantsker (✉)
Department of Chemistry, University of Illinois at Chicago, Chicago, IL, USA
e-mail: dantsker@uic.edu

© Springer Nature Switzerland AG 2019
A. Rosenhouse-Dantsker, A. N. Bukiya (eds.), *Direct Mechanisms in Cholesterol Modulation of Protein Function*, Advances in Experimental Medicine and Biology 1135, https://doi.org/10.1007/978-3-030-14265-0_7

119

1 Introduction

In recent years, it has been demonstrated that a growing number of ion channels are modulated by cholesterol, a major lipid component of the plasma membrane (e.g. reviewed in [1–3]). In most cases, an inverse relationship was observed between cholesterol levels and channel activity. Accordingly, a decrease in cholesterol content resulted in an increase in channel function whereas an increase in cholesterol levels led to a decrease in channel activity. Such an inverse relationship has been observed for several inwardly rectifying potassium channels (e.g. Kir1.1, Kir2.1, Kir3.1*(GIRK1*), the homomerically active F137S pore mutant of Kir3.1 [4], and Kir6.2Δ36, the C-terminal truncation mutant of Kir6.2 that renders these channels active as homomers in the absence of sulfonylurea receptor (SUR) subunits [5]) [6–8] (Fig. 1), for voltage gated potassium, sodium and calcium channels [9–14], and for volume-regulated anion channels [15]. In contrast, very few channels exhibited a direct relationship between cholesterol levels and channel activity in which increased levels of cholesterol led to increased channel function whereas decreased levels of cholesterol resulted in a decrease in channel function. These included the epithelial sodium channels (eNaC) [16, 17], the transient receptor potential canonical channel TRPC1 [18], and the Ca^{2+}-permeable stretch-activated cation channels (SACs) [19]. In addition, we have shown that the heterotetrameric Kir3 channels in atrial myocytes (Kir3.1/Kir3.4) [8] and in hippocampal neurons from the CA1 region (Kir3.1/Kir3.2/Kir3.3) [20] were also up-regulated by cholesterol. This effect was mimicked by pore mutants of the homomeric Kir3.2 (GIRK2) and Kir3.4 (GIRK4) channels that constitute the primary Kir3 subunits in the brain and heart, respectively (Fig. 1). Specifically, both Kir3.2^ (GIRK2^) and Kir3.4* (GIRK4*) were up-regulated by cholesterol [8, 20]. Kir3.2^ and Kir3.4* are the Kir3.2_E152D

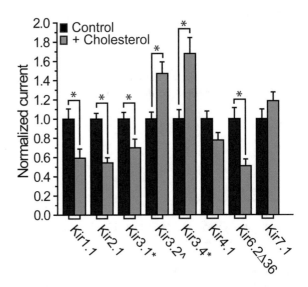

Fig. 1 Effect of cholesterol enrichment on Kir channels. Whole cell basal currents of control and cholesterol-enriched *Xenopus oocytes* injected with representative homotetrameric Kir channels ($n = 10$–64). Significant difference is indicated by an asterisk ($*p \leq 0.05$). The figure combines data that were originally published in [6, 8, 20]

and Kir3.4_S143T pore mutants of Kir3.2 and Kir3.4, respectively, that enhance the activity of these homomeric channels [4, 21, 22].

The effect of cholesterol on ion channels was initially attributed to the effect of its rigid fused ring system on the plasma membrane lipid bilayers (reviewed by [3]). This notion was supported by observations showing that alterations in cholesterol levels lead to changes in the physical properties of lipid bilayers including their rigidity, fluidity, and thickness. It was thus proposed that these modifications in bilayer properties result in a hydrophobic mismatch between the lipid bilayer and the transmembrane domains of membrane proteins, thereby affecting their function [23, 24].

However, in recent years, a growing number of crystallographic structures have demonstrated that cholesterol can bind to transmembrane proteins, suggesting that it can affect protein function directly [reviewed in 25]. These include a variety of G-protein coupled receptors, transporters, and the sodium potassium pump, among others [e.g. 26–34]. Furthermore, there is growing evidence that suggests that cholesterol may affect the function of ion channels directly by binding to the channels [3, 35–40]. While ion channels complexed with cholesterol have yet to be crystallized, a cryo-electron microscopy structure of TRPM2 in complex with cholesterol has been recently determined at a 3.07 Å resolution demonstrating that cholesterol may bind directly to ion channels [41]. The concept that cholesterol may bind to ion channels can be traced to earlier studies on several ion channels (e.g. Kir, BK, nicotinic acetylcholine receptor, TRPV1) that demonstrated lack of correlation between the effect of different sterols on membrane properties and channel function [42–47]. Specifically, studies on the specificity of the nicotinic acetylcholine receptor sterol activation indicated that it was independent of sterol features that affect bilayer properties [43]. Similarly, no correlation was observed between the effect of different sterols on membrane fluidity and the function of the bacterial KirBac1.1 channel [44]. Consistent with these observations, studies on the large conductance Ca^{2+} and voltage-gated K^+ (BK) channel suggested that an increase in bilayer lateral stress was unlikely to underlie the differential effect of cholesterol and other sterols on BK channels function [46]. Notably, unlike cholesterol, its enantiomer, ent-cholesterol [48], did not have any effect on KirBac1.1, Kir2.1, and BK channels further supporting the notion that cholesterol efficacy requires stereospecific sterol recognition by the channel protein [45, 46].

More recently, computational studies have identified putative cholesterol binding sites in several ion channels including the nicotinic acetylcholine receptor and the TRPV channel [47, 49, 50]. Additionally, we have identified putative cholesterol binding sites in Kir2.1, Kir3.2 and Kir3.4 [20, 51, 52]. Functional studies have corroborated the importance of key residues within these binding sites to the cholesterol sensitivity of the channels.

One of the perplexing observations in terms of the effect of cholesterol on ion channels is the differential effect of cholesterol on different inwardly rectifying potassium channels. The inwardly rectifying potassium channel family includes fifteen members that have been classified into seven subfamilies (Kir1-7). Kir channels regulate multiple cellular functions in a variety of tissues including membrane excitability in cardiomyocytes, muscle cells and neurons, heart rate, vascular tone, insulin release and salt flow across epithelia (reviewed by e.g. [53–57]). Consequently, the effect of alterations in cholesterol levels on Kir function can have significant physiological implications. Figure 1 summarizes the effect of cholesterol enrichment on the activity of representative homomeric Kir channels from all Kir sub-families except for Kir5.1, which is not functional as a homomer. As noted above and evident in Fig. 1, the impact (up-regulation versus down-regulation) of cholesterol varies among the different Kir channels. Yet, the structure of all Kir channels is very similar. They are all composed of four subunits that are each formed from two membrane spanning helices connected by an extracellular domain and a pore helix, and cytosolic N- and C- termini [53, 58–65]. Moreover, the homology among Kir channels is relatively high with an identity that ranges between ~28% (between Kir7.1 and Kir6.1 or Kir6.2) and ~70% (between Kir2.1 and Kir2.2, and between Kir3.2 and Kir3.4). In particular, the identity between Kir2.1 and Kir3.2 or Kir3.4 is ~45%. It is therefore unsurprising that our recent studies of Kir2.1, Kir3.2 and Kir3.4 have identified putative cholesterol binding sites in equivalent trans-membrane domains of these three channels. However, the structural basis for the opposite impact of cholesterol on Kir2.1 versus Kir3.2^ and Kir3.4* remains puz-zling. Thus, focusing on Kir2.1 and the Kir3 channels, Kir3.2^ and Kir3.4*, we compared the residues that participate in the putative cholesterol binding pockets in these three channels.

2   Putative Cholesterol Binding Sites in Kir2.1, Kir3.2 and Kir3.4

2.1   Cholesterol Binding Sites in Kir2.1

The four Kir2 channels, Kir2.x where $x = 1$–4, were the first inwardly rectifying potassium channels shown to be modulated by cholesterol. All four channels were suppressed by cholesterol, albeit to different degrees [6, 7, 42, 66].

In a recent extensive computational-experimental study, we identified two puta-tive cholesterol binding sites in the transmembrane domain of Kir2.1 (Fig. 2) [52]. Specifically, molecular docking followed by molecular dynamics simulations pre-dicted that cholesterol may bind to the channel either at a site located at the center of the transmembrane domain or at a site at the interface of the transmembrane and cytosolic domains. In agreement with these predictions, functional studies identi-fied several residues within the two putative binding sites whose mutation signifi-cantly affected the sensitivity of the channel to cholesterol.

Among the residues whose mutation abrogated or significantly reduced the sen-sitivity of the channel to cholesterol, we identified 10 hydrophobic aliphatic resi-dues (A, I, L, V, and M), 2 aromatic residues (Y, F), and 3 small polar residues (C, S) [52]. Whereas the majority of the hydrophobic residues and one of the small

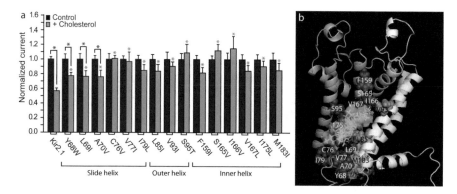

Fig. 2 Putative cholesterol binding sites in Kir2.1 are located in the transmembrane domain of the channel. (**a**) Whole-cell basal currents recorded in *Xenopus oocytes* at −80 mV showing the effect of cholesterol enrichment on Kir2.1 WT and on the mutants listed on the x-axis of the figure ($n = 9–90$). Error bars, S.E. Significant difference is indicated by an asterisk (*, $p ≤ 0.05$). Blue asterisks indicate statistically significant difference with respect to the effect of cholesterol on the WT construct. (**b**) Ribbon representation of the transmembrane and extracellular domains of two adjacent subunits of Kir2.1 (in gray and white) depicting in ball representation (in brick red in one subunit, and salmon pink in the second subunit) the residues whose mutation affected the sensitivity of the channel to cholesterol. Also shown are the locations of the cholesterol molecules at the two putative binding regions (cyan sticks and surface representations). The figure is based on data that was originally published in [52]

polar residue (S) were located within van der Waals interaction range from at least one of the cholesterol molecules in the model, backbone atoms of a few of the hydrophobic residues were within a hydrogen bonding distance from the hydroxyl group of the cholesterol molecule. In contrast, both aromatic residues (Y68 and F159) were not interacting directly with any of the cholesterol molecules. Additionally, one hydrophobic residue (I79) and two of the small polar residues (C76 and S165) were also not located within interaction distance from either of the cholesterol putative binding sites. The mutation of these five residues likely affected the sensitivity of the channel to cholesterol indirectly through residues that directly interacted with the cholesterol molecule.

The 10 cholesterol-interacting residues whose mutation affected the sensitivity of the channel to cholesterol did not form a continuous chain. The mutation of several other residues that were located within interaction distance from one or both of the cholesterol molecules did not affect the sensitivity of the channel to cholesterol (Fig. 3) [52]. These included several hydrophobic aliphatic residues and aromatic residues among others. One possibility is that these residues were not critical for cholesterol recognition and/or binding. Alternatively, it is also possible that some of the mild mutations made, which were designed to minimize the likelihood of loss of function and non-specific effects, could be tolerated in these particular positions.

Both putative binding sites were located in distinct pockets in between transmembrane α-helices of two adjacent subunits. Consequently, membrane phospholipids were occluded from binding to these non-annular sites [36] in molecular

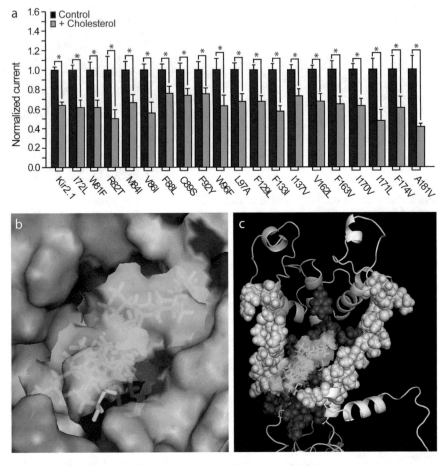

Fig. 3 Transmembrane residues that do not affect the sensitivity of the Kir2.1 to cholesterol engulf the putative 'principal' cholesterol binding site. (**a**) Whole-cell basal currents recorded in *Xenopus oocytes* at −80 mV showing the effect of cholesterol enrichment on Kir2.1 WT and on the mutants listed on the x-axis of the figure ($n = 7$–90). Error bars, S.E. Significant difference is indicated by an asterisk (*, $p \leq 0.05$). (**b**) Surface representation of the putative 'principal' cholesterol binding region in Kir2.1 showing proximal residues whose mutation affected the sensitivity of the channel to cholesterol (in brick red) along with residues whose mutation did not affect the sensitivity of the channel to cholesterol (in wheat yellow). Putative orientations of the cholesterol molecule within this binding region are also shown (cyan sticks and surface representations). (**c**) Ribbon representation of the transmembrane and extracellular domains of two adjacent subunits of Kir2.1 (in gray and white) depicting cholesterol residues in ball representation (in brick red) whose mutation affected the sensitivity of the channel to cholesterol as well as residues whose mutation did not affect the sensitivity of the channel to cholesterol (in wheat yellow). Also shown are the locations of the cholesterol molecules at the two putative binding regions (cyan sticks and surface representations). The figure is based on data that were originally published in [52]

dynamics simulations [52]. Notably, outside the two putative binding pockets, the mutation of multiple transmembrane residues at annular sites of the channel that were accessible to membrane phospholipids did not affect the sensitivity of the channel to cholesterol (see examples in Fig. 3).

Computational analysis of the binding enthalpy and free energy, which are indicators of the binding energy and binding affinity of cholesterol to the channel, suggested that the interaction between the cholesterol molecule and the binding pocket located at the center of the transmembrane domain may be stronger than the interaction of cholesterol with the pocket located at the interface of the transmembrane and cytosolic domains of the channel [52]. Yet, mutations of residues in both binding pockets abrogated the sensitivity of the channel to cholesterol. This suggested that the binding pocket at the interface between the transmembrane and cytosolic domains is a short-lived 'transient' cholesterol binding site to which cholesterol binds on its way to the more stable 'principal' binding site at the center of the transmembrane domain.

2.2 Cholesterol Binding Sites in Kir3.2

In a more recent study, we determined that cholesterol up-regulates Kir3 channels expressed in the CA1 region of the hippocampus [20]. Within this region of the brain, Kir3.1, Kir3.2a, Kir3.2c, and Kir3.3 combine to form functional heterotetramers [67–69]. Among these Kir3 subunits, only Kir3.2 can form functional homotetramers [70–72]. Notably, Kir3.2^ homotetramers were up-regulated by cholesterol similarly to the native hippocampal channels [20].

We thus employed a computational-experimental approach to determine whether despite the opposite impact of cholesterol on Kir2.1 (down-regulated) and Kir3.2^ (up-regulated), cholesterol binds to similar regions in the two inwardly rectifying potassium channels. This was achieved in two stages (Fig. 4) [20]. First, unbiased computational studies were utilized to search for potential cholesterol binding sites in Kir3.2^. These led to the identification of two possible binding sites in two transmembrane regions that were similar to the cholesterol binding regions in Kir2.1 [20, 52]. Specifically, one was located at the center of the transmembrane domain and the other, at the interface between the transmembrane and cytosolic domains of the channel (Fig. 4c). Second, to test the predictions of the computational analysis, four different types of residues were defined based on their potential effect on the sensitivity of the channel to cholesterol. The effect of the mutation of representative residues of each of these types of residues on the sensitivity of the channel to cholesterol was then determined experimentally (Fig. 4b) [20]. The residues interrogated were: (1) V99 and L174 in one binding site and V101 and V183 in the second binding site. According to the computational model, these residues were located at the bottom of each of the two putative binding sites, and as such, directly interacted with the cholesterol molecule, and were thus expected to play a key role in the sensitivity of the channel to cholesterol; (2) L86, a residue that could potentially affect the sensitivity

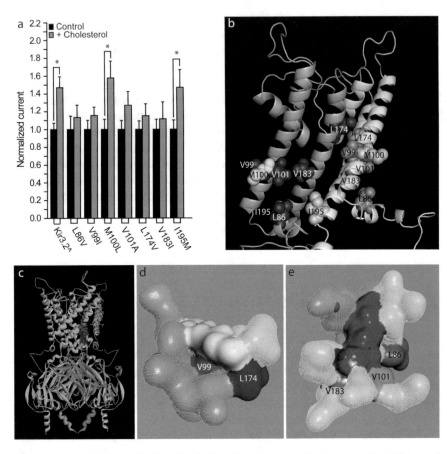

Fig. 4 Putative cholesterol binding sites in Kir3.2^ are located in the transmembrane domain of the channel. (**a**) Whole cell basal currents recorded in *Xenopus oocytes* at −80 mV showing the effect of cholesterol enrichment on Kir3.2^ and L86V, V99I, M100L, V101A, L174V, V183I, and I195M ($n = 11$–23). Significant difference is indicated by an asterisk (*, $p \leq 0.05$). (**b**) Ribbon representation of the transmembrane and extracellular domains of two adjacent subunits of Kir3.2^ (in gray and white) depicting in ball representation (in brick red in one subunit, and salmon pink in the second subunit) the residues whose mutation affected the sensitivity of the channel to cholesterol. Also shown in ball representation are residues whose mutation did not affect the sensitivity of the channel to cholesterol (in yellow in one subunit, and wheat yellow in the second subunit). (**c**) Location of the cholesterol molecule (yellow and magenta) at two putative cholesterol-binding sites of Kir3.2^ between two adjacent subunits of the channel (cyan and violet) as obtained from molecular docking. (**d**) Surface representation depicting the residues that surround the cholesterol molecule (yellow) in site 1 including V99 and L174. (**e**) Surface representation depicting the residues that surround the cholesterol molecule (magenta) in site 2 including L86, V101, and V183. Figures **a**, **c**–**e** were originally published in [52]. Figure **b** is based on data that were originally published in [52]

of the channel to cholesterol indirectly via F186, a directly interacting residue; (3) I195, a residue located at a mutation-tolerant position in proximity to the cholesterol molecule where it was not tightly-packed, and was thus unlikely to play a critical role in the sensitivity of Kir3.2 to cholesterol; (4) M100, a residue that faced away from both putative binding sites, and was thus not expected to affect the sensitivity of the channel to cholesterol. As shown in Fig. 4b, there was clear correspondence between the computational predictions obtained and the functional data. Specifically, five of the seven mutants tested were insensitive to cholesterol. These constructs included single mutations of residues that either directly interacted with the cholesterol molecule in each of the potential binding sites (type (1)), or indirectly affected the sensitivity of the channel to cholesterol via a directly interacting residue (type (2)) (Fig. 4d, e). Mutation of residues of types (3) and (4) did not affect the sensitivity of the channel to cholesterol as predicted by the computational model (Fig. 4a, b), further supporting the notion that Kir3.2 possesses two putative cholesterol binding sites in similar regions to the regions in which the 'principal' and 'transient' sites in Kir2.1 were located.

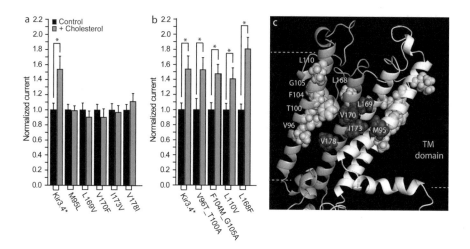

Fig. 5 The effect of cholesterol on Kir3.4* depends on residues in the transmembrane domain of the channel. (**a, b**) Whole-cell basal currents recorded in *Xenopus oocytes* at − 80 mV showing the effect of cholesterol enrichment on Kir3.4* and (**a**) M95V, L169V, V170F, I173V, and V178I (n-13-21), and (**b**) V196T_T100A, F104M_G105A, L110V, and L168F ($n = 11$–21). Significant difference is indicated by an asterisk (*$p \leq 0.05$). (**c**) A ribbon representation of two adjacent subunits of the Kir3.4* channel showing the transmembrane residues (in ball representation) whose effect on cholesterol sensitivity of the channel was tested in mutagenesis studies displayed in Fig. 4a, b. The location of the residues whose mutation abrogated cholesterol sensitivity of the channel is shown (in brick red) in one subunit (gray) for residues L169, V170, I173, and V178, and in the adjacent subunit (white) for M95. The location of the residues whose mutation did not affect the sensitivity of the channel to cholesterol is depicted in duplicate in two adjacent subunits of the channel (in wheat yellow). The figure is based on data that were originally published in [51]

2.3   Cholesterol Binding Site in Kir3.4

Next, we investigated whether cholesterol binds to Kir3.4* (Fig. 5) [51]. In contrast to the brain, in which the predominant Kir3 subunits are Kir3.1, Kir3.2, and Kir3.3, in the heart, the prevalent Kir3 subunits are Kir3.1 and Kir3.4. In an earlier study, we have demonstrated that cholesterol upregulates atrial Kir3 channels [8]. Similarly, the homomeric Kir3.4* channels were also up-regulated by cholesterol (Fig. 1) [6, 8]. Thus, in a recent study that demonstrated that cholesterol acted synergistically with $PI(4,5)P_2$ to activate Kir3.2, we also examined the relative location of the putative cholesterol and $PI(4,5)P_2$ binding sites [51]. While the focus of the study was different from the studies on Kir2.1 [52] and Kir3.2^ [20], it led to several conclusions with respect to the binding of cholesterol to Kir3.4*. First, in contrast to $PI(4,5)P_2$ that was shown to bind to a cluster of primarily positively charged residues within the cytosolic domain of inwardly rectifying potassium channels at the interface with the transmembrane domain [e.g. 3, 73–75], the putative cholesterol binding site was located in the transmembrane domain [51]. Focusing on the 'principal' binding region of cholesterol, we identified five hydrophobic aliphatic residues (M, L, V, I) in a non-annular surface of the channel whose mutation abrogated the sensitivity of the channel to cholesterol (Fig. 5a, c). These five residues were split between the transmembrane helices of two adjacent subunits of the channel protein. In contrast, the mutation of six annular residues (G, L, V, F) that we tested had no effect on the sensitivity of Kir3.4* to cholesterol (Fig. 5b). Two sets of these six annular residues (from two different subunits) engulfed the non-annular cluster of the residues that were found to play a key role in the modulation of the channel by cholesterol (Fig. 5c). Thus, similarly to Kir2.1 and Kir3.2^, these data suggested that also in Kir3.4*, cholesterol binds to a similar non-annular hydrophobic 'principal' region of the transmembrane domain of the channel. Further studies are required to determine whether Kir3.4* also possesses a 'transient' cholesterol binding site.

3   Common Characteristics of Putative Cholesterol Binding Pockets in Kir2.1, Kir3.2^ and Kir3.4*: Implications for Cholesterol Binding to Kir Channels

In summary, the studies of Kir2.1, Kir3.2^ and Kir3.4* described above identified putative non-annular cholesterol binding sites in between α-helices within the transmembrane domain of each of the three channels [20, 51, 52]. The notion that cholesterol may bind to membrane proteins in between different α-helices within the transmembrane domain has been previously demonstrated for a variety of G-protein coupled receptors [76] including, for example, the β-adrenergic receptor [77], the proton pumping Rhodopsin receptor ARII [27], the μ-opioid receptor [78], the A2A adenosine receptor [79], the dopamine receptor [32], and the 5-HT2B/ERG receptor [80]. Interestingly, however, in all three Kir channels studied, the putative cholesterol binding sites were not located in between α-helices of the same subunit, but rather in between α-helices of adjacent channel subunits.

Another common feature of the putative cholesterol binding sites identified in Kir2.1, Kir3.2^ and Kir3.4* is the types of residues forming these binding sites. Being a primarily hydrophobic molecule, cholesterol binding requires a hydrophobic environment [25]. This is evident in the growing number of crystallographic structures of different proteins complexed with cholesterol available in the Research Collaboratory for Structural Bioinformatics Protein Data Bank (RCSB PDB) [e.g. 26–34, 73–76] as well as in several cholesterol binding motifs suggested in earlier studies. First, in studies of the peripheral-type benzodiazepine receptor, a transmembrane protein that mediates the translocation of cholesterol, the sequence -(L/V)X_{1-5}YX$_{1-5}$(R/K)- (where X_{1-5} represents 1–5 residues of any amino acid) was proposed to act as a cholesterol recognition amino acid consensus (CRAC) binding motif [81]. The CRAC motif was subsequently also identified in ion channels including the BK channels [82], and the nicotinic acetylcholine receptor (nAChR) [50]. More recently, an inverted CRAC sequence, (R/K)X_{1-5}(Y/F)X$_{1-5}$(L/V), has been suggested to be responsible for cholesterol interactions with AChR [50]. This motif, named CARC, was identified in the transmembrane domain of the TRPV1 channel [47]. Both CRAC and CARC motifs involved in addition to a hydrophobic residue, also an aromatic residue and a positively charged residue. Whereas the aromatic residue was expected to interact with the ring structure of cholesterol via stacking interactions, the positively charged residue was expected to interact with the hydroxyl group of the cholesterol molecule [25]. Similarly, the cholesterol consensus motif (CCM) [77] that was based on the complexed crystal structure of the β-adrenergic receptor with cholesterol included the same types of residues in two adjacent helices of the G-protein coupled receptor: (W/Y)(I/V/L)(K/R) on one helix and (F/Y/R) on a second helix [26, 77].

In alignment with the requirement for a hydrophobic environment, the putative cholesterol binding sites in Kir2.1, Kir3.2^ and Kir3.4* were primarily formed by hydrophobic residues (Figs. 2, 3, 4, and 5) [20, 51, 52]. We also identified aromatic residues in proximity to the putative cholesterol binding site in Kir2.1 whose mutation abrogated or significantly reduced the sensitivity of the channel to cholesterol (Fig. 2) [52]. Furthermore, although the effect of mutations of aromatic residues within the putative cholesterol binding sites of Kir3.2^ and Kir3.4* were not tested experimentally, aromatic residues were located within these regions of the channel, and could potentially play a critical role in cholesterol modulation of the channels. In contrast, positively charged residues that could potentially interact with the hydroxyl group of the cholesterol molecule were not identified within the putative binding sites of these channels. Instead, the hydroxyl group could potentially interact with backbone atoms of other types of residues (including hydrophobic residues) as occurred in a variety of proteins to which cholesterol has been shown to bind [25].

While our studies have primarily focused on Kir2.1, Kir3.2^ and Kir3.4*, we have also shown that other Kir channels are also modulated by cholesterol (Fig. 1).

The studies on Kir2.1, Kir3.2^ and Kir3.4* described above [20, 51, 52] suggest that independent of the impact of cholesterol on the channel (up-regulated versus down-regulated), putative cholesterol binding sites in Kir channels are (1) non-annular transmembrane sites [36]; (2) located between α-helices of two adjacent channel subunits; and (3) involve hydrophobic and aromatic residues. Further studies are required to determine whether cholesterol binding sites in other Kir channels share the same characteristics.

4   Differences Between the Putative Binding Sites in Kir2.1, Kir3.2^ and Kir3.4*: Implications for the Impact of Cholesterol on Ion Channels

While the general characteristics of the putative cholesterol binding sites in Kir2.1, Kir3.2^ and Kir3.4* are similar, there are some intriguing differences between them. Figure 6a depicts the aligned sequences of the slide helices, inner helices and outer helices of the three channels. Residues tested experimentally for their potential effect on the sensitivity of the channels to cholesterol are highlighted in the figure [20, 51, 52, 83]. Taken together with Figs. 2, 3, 4, and 5 showing the location of the putative binding sites in Kir2.1, Kir3.2^ and Kir3.4*, it seems that the 'principal' binding site is shifted towards the extracellular domain in the two Kir3 channels compared to its location in Kir2.1 (Fig. 6b). This is especially evident in the inner helix where S165 is the closest residue to the extracellular domain in the 'principal' binding site whose mutation affected Kir2.1 sensitivity to cholesterol. In contrast, mutation of V162, which is located about a helical turn extracellular to S165, did not have any effect on the sensitivity of the channel to cholesterol, and according to the computational analysis, is not a part of the cholesterol binding site [52]. In Kir3.2^, on the other hand, mutation of the equivalent residue, I174, abrogated the sensitivity of Kir3.2^ to cholesterol and according to the computational analysis, is a part of the putative binding site of cholesterol in the channel [20]. Similarly, mutation of the equivalent residue in Kir3.4*, L169, abrogated the sensitivity of Kir3.4* to cholesterol [51].

Some shift in the position of the 'transient' putative binding site was also observed in Kir3.2^ compared to Kir2.1. As a result of these shifts in both putative binding sites, the majority of the residues that formed cholesterol binding sites in Kir2.1 were not located at equivalent positions to those of the residues that formed cholesterol binding sites in the Kir3 channels, and only very few mutations of pairs of equivalent residues in Kir2.1 and Kir3.2^ or Kir3.4* resulted in the same effect on cholesterol sensitivity (Fig. 6a) [20, 51, 52].

Could this shift in the location of the putative 'principal' cholesterol binding sites in Kir3.2^ and Kir3.4* compared to Kir2.1 underlie the opposite impact of cholesterol on the channels? Further studies are required to address this question. It is intriguing, however, that whereas the 'principal' cholesterol binding site in Kir2.1

a

Slide helix
Kir2.1   68 YLADIFTTCVDI 79   ■ Effect
Kir3.2^   78 YLTDIFTTLVDL 89
Kir3.4*   73 YLSDLFTTLVDL 84   ░ No effect

Outer helix
Kir2.1   83 WMLVIFCLAFVLSWLFFGCVFWLIA 107
Kir3.2^   93 FNLLIFVMVYTVTWLFFGMIWWLIA 117
Kir3.4*   88 FNLLVFTMVYTVTWLFFGFIWWLIA 112

Inner helix
Kir2.1   157 AVFMVVFQSIVGCIIDAFIIGAVMAKMAK 185
Kir3.2^ 169 GIILLLIQSVLGSIVNAFMVGCMFVKISQ 197
Kir3.4* 164 GIILLLVQAILGSIVNAFMVGCMFVKISQ 192

b

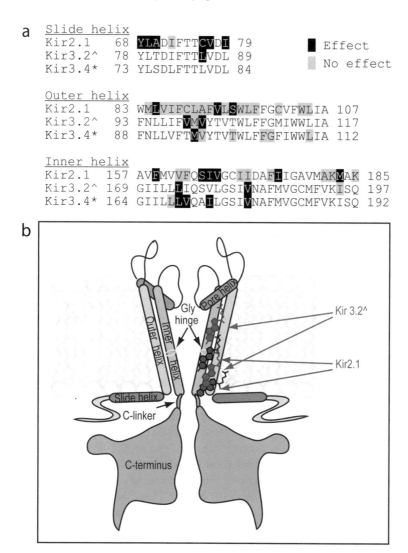

Fig. 6 Differences and similarities in the locations of the putative cholesterol binding sites in Kir2.1, Kir3.2^, and Kir3.4*. (**a**) Sequence alignment of the slide helices, outer helices and inner helices of Kir2.1, Kir3.2^, and Kir3.4* showing residues tested for their effect on cholesterol sensitivity. Highlighted in gray are residues whose mutation did not affect the sensitivity of the channel to cholesterol. Highlighted in black are residues whose mutation significantly affected the sensitivity of the channel to cholesterol. The residues highlighted are based on experimental data from [20, 51, 52, 83]. (**b**) schematic model illustrating the location of the two cholesterol putative binding regions in Kir2.1 and Kir3.2^ along with labeling of key channel regions. Note that for purposes of clarity, the model shows the cholesterol molecules next to one of the two adjacent channel subunits with which they are predicted to interact

overlaps with the hinge region of the inner transmembrane helix of the channel, the putative 'principal' cholesterol binding sites in the two Kir3 channels are located extracellular to this region. Comparison of the crystal structures of two related bacterial channels, KcsA and MthK, suggested that a highly conserved central glycine of the inner transmembrane helix may play the role of a gating hinge in the two channels [84]. This notion was further supported in a study showing that the corresponding central glycine in Kir3.4* played a central role in channel gating. It was thus suggested that the flexibility of the glycine is critical for channel gating at the helix bundle crossing [85]. Our further analysis of neighboring residues in Kir3.4* suggested that hinging occurs not at the glycine itself but at the residue that immediately precedes it [86]. Thus, with the putative cholesterol 'principal' binding site in Kir2.1 overlapping with this region, it is plausible that cholesterol binding interferes with the hinging motion of the inner helix, thereby stabilizing Kir2.1 in the closed state [52]. In contrast, the putative binding of cholesterol in Kir3.2^ and Kir3.4* just past the hinge region of the inner helix towards the extracellular domain may impart an opposite effect on channel function.

Aside from the differences between Kir2.1 and the two Kir3 channels, there were surprising differences between the effect of mutations of specific outer helix residues of the two Kir3 channels on their sensitivity to cholesterol. Whereas mutations of equivalent residues in the inner helix of the two channels had the same effect on the sensitivity of the channels to cholesterol, mutations of two pairs of outer helix residues had the opposite effect [87]. Specifically, whereas the M100L mutation in Kir3.2^ had no effect on the sensitivity of the channel to cholesterol, the M95L mutation in Kir3.4* abrogated its sensitivity to cholesterol. Conversely, mutation of V101 in Kir3.2^ abrogated its sensitivity to cholesterol whereas mutation of the equivalent residue in Kir3.4*, V96, had no effect on the sensitivity of the channel to cholesterol. Accordingly, the putative 'principal' cholesterol binding sites in Kir3.2^ and Kir3.4* were not entirely formed from equivalent channel residues. Moreover, as a part of the α-helical structure of the transmembrane domain of the channels, these consecutive residues (M100, V101 in Kir3.2^ and M95, V96 in Kir3.4*) face in different directions as each amino acid corresponds to a ~100° turn in the helix. Thus, in the absence of a crystal structure of the transmembrane domain of Kir3.4, this raises the question whether the outer transmembrane helices in these two channels are oriented in a slightly different manner.

Recently, the cryo-electron microscopy structure of the TRPM2 channel revealed a cholesterol binding site in the transmembrane domain of the channel (Fig. 7a) [41]. Similarly to the putative cholesterol binding sites in Kir channels [20, 51, 52], the binding site of cholesterol in TRPM2 [41] was located in between α-helices of two adjacent channel subunits. Also, the types of TRPM2 residues interacting with cholesterol were similar to the residues involved in cholesterol interactions in Kir channels, and included hydrophobic and aromatic residues (Fig. 7b). Notably, the binding site of cholesterol in TRPM2 was located close to the extracellular domain. In contrast, another member of the TRP superfamily of channels, TRPV1, was shown to possess a CARC motif at the center of the S5 transmembrane helix close to the intracellular domain [47]. Mutations of the R579, F582, and L585 residues

Fig. 7 The binding site of cholesterol in TRPM2 is located in the transmembrane domain of the channel. (**a**) A ribbon representation of two adjacent subunits (in teal and violet) of the TRPM2 channel showing the location of the binding site of cholesterol (in yellow) within the transmembrane domain of the channel close to the extracellular domain. The figure is based on PDB ID 6CO7. (**b**) Surface representation depicting the residues that surround the cholesterol molecule (in yellow) in TRPM2

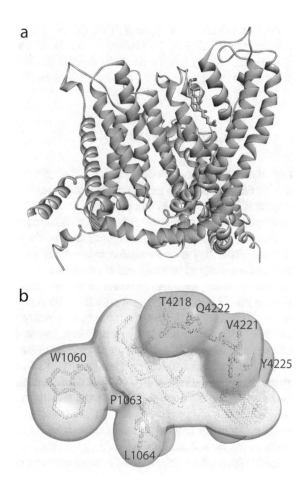

within this cholesterol binding motif abrogated or significantly reduced the sensitivity of the channel to cholesterol. The location of the CARC motif was similar to the location of the principal putative binding site in Kir2.1. In particular, the position of the L585 residue in the S5 helix of TRPV1 corresponds to the position of V93 in the outer helix of Kir2.1 [52]. Interestingly, similarly to Kir2.1, several genetic variants of TRPV1 were down-regulated by cholesterol [47]. In contrast, it has been shown that a different member of the TRP superfamily, TRPC1, is up-regulated by cholesterol [18]. This suggests that as in the case of Kir channels, the impact of cholesterol varies among TRP channels as well. As subtle differences in the sequence of Kir channels have a substantial effect on the impact of cholesterol on the channels, it is difficult to predict how TRPM2 will respond to alterations in cholesterol levels. However, as both Kir3.2^ and Kir3.4* possess a putative cholesterol binding site close to the extracellular domain and are both up-regulated by cholesterol, the possibility that TRPM2 would also be up-regulated by cholesterol is intriguing given the location of the binding site of cholesterol in this channel.

Further experiments are required to test this possibility, and address the question of whether the location of the binding site of cholesterol within the transmembrane domain (i.e. closer to the extracellular domain or closer to the intracellular domain) plays a role in determining the impact of cholesterol on channel function.

5 Concluding Remarks

Our studies on the modulation of Kir2.1, Kir3.2^ and Kir3.4* [20, 51, 52] suggest that cholesterol binds to the transmembrane domain of all three channels albeit at slightly different locations and/or orientations. Specifically, the studies suggest that independent of the impact of cholesterol on the channel (up-regulated versus down-regulated), putative cholesterol binding sites in Kir channels share several characteristics. First, they are non-annular transmembrane sites. Second, they are located between α-helices of two adjacent channel subunits. Third, they involve hydrophobic and aromatic residues. However, whereas the putative cholesterol 'principal' binding site in Kir2.1 overlapped with the central glycine hinge region of the inner helix of the channel, the putative 'principal' binding sites of cholesterol in Kir3.2^ and Kir3.4* were located just past the hinge region of the inner helix towards the extracellular domain. These differences may be at the core of the opposite impact of cholesterol on Kir2.1 (down-regulated) versus Kir3.2^ and Kir3.4* (up-regulated). Furthermore, mutations of two pairs of equivalent outer helix residues in Kir3.2^ and Kir3.4* had the opposite effect on the sensitivity of the channels to cholesterol suggesting that the putative 'principal' cholesterol binding sites in these two Kir3 channels were not entirely formed from equivalent channel residues. These differences in the cholesterol binding site-forming residues of Kir3.2^ and Kir3.4* may originate from subtle structural differences between these highly homologous channels (e.g. variations in helical orientation).

Notably, the studies described in this chapter focused on one channel at a time, and had different goals. Consequently, different sets of residues in Kir2.1, Kir3.2^ and Kir3.4* were tested for their role in the sensitivity of the channels to cholesterol. Thus, further studies are required to systematically establish the observations described above, address lingering questions, and elucidate the structural basis of the mechanisms that underlie the modulation of these channels by cholesterol.

References

1. Maguy A, Hebert TE, Nattel S. Involvement of lipid rafts and caveolae in cardiac ion channel function. Cardiovasc Res. 2006;69:798–807.
2. Levitan I, Fang Y, Rosenhouse-Dantsker A, Romanenko V. Cholesterol and ion channels. Subcell Biochem. 2010;51:509–49.
3. Rosenhouse-Dantsker A, Mehta D, Levitan I. Regulation of ion channels by membrane lipids. Compr Physiol. 2012;2:3168.

4. Chan KW, Sui JL, Vivaudou M, Logothetis DE. Control of channel activity through a unique amino acid residue of a G protein-gated inwardly rectifying K⁺ channel subunit. Proc Natl Acad Sci U S A. 1996;93:14193–8.

5. Tucker SJ, Gribble FM, Zhao C, Trapp S, Ashcroft FM. Truncation of Kir6.2 produces ATP-sensitive K⁺ channels in the absence of the sulphonylurea receptor. Nature. 1997;387:179–83.

6. Rosenhouse-Dantsker A, Leal-Pinto E, Logothetis DE, Levitan I. Comparative analysis of cholesterol sensitivity of Kir channels: Role of the CD loop. Channels. 2010;4:63–6.

7. Romanenko VG, Fang Y, Byfield F, Travis AJ, Vandenberg CA, Rothblat GH, Levitan I. Cholesterol sensitivity and lipid raft targeting of Kir2.1 channels. Biophys J. 2004;87:3850–61.

8. Deng W, Bukiya AN, Rodríguez-Menchaca AA, Zhang Z, Baumgarten CM, Logothetis DE, Levitan I, Rosenhouse-Dantsker A. Hypercholesterolemia induces up-regulation of K_ACh cardiac currents via a mechanism independent of phosphatidylinositol 4,5-bisphosphate and Gβγ. J Biol Chem. 2012;287:4925–35.

9. Balajthy A, Hajdu P, Panyi G, Varga Z. Sterol regulation of voltage-gated K⁺ channels. Curr Top Membr. 2017;80:255–92.

10. Bolotina V, Omelyanenko V, Heyes B, Ryan U, Bregestovski P. Variations of membrane cholesterol alter the kinetics of Ca²⁺-dependent K⁺ channels and membrane fluidity in vascular smooth muscle cells. Pflugers Arch. 1989;415:262–8.

11. Dopico AM, Bukiya AN, Singh AK. Large conductance, calcium- and voltage-gated potassium (BK) channels: regulation by cholesterol. Pharmacol Ther. 2012;135:133–50.

12. Wu CC, Su MJ, Chi JF, Chen WJ, Hsu HC, Lee YT. The effect of hypercholesterolemia on the sodium inward currents in cardiac myocyte. J Mol Cell Cardiol. 1995;27:1263–9.

13. Amsalem M, Poilbout C, Ferracci G, Delmas P, Padilla F. Membrane cholesterol depletion as a trigger of Nav1.9 channel-mediated inflammatory pain. EMBO J. 2018;37(8):e97349.

14. Toselli M, Biella G, Taglietti V, Cazzaniga E, Parenti M. Caveolin-1 expression and membrane cholesterol content modulate N-type calcium channel activity in NG108-15 cells. Biophys J. 2005;89:2443–57.

15. Levitan I, Christian AE, Tulenko TN, Rothblat GH. (2000) Membrane cholesterol content modulates activation of volume-regulated anion current in bovine endothelial cells. J Gen Physiol. 2000;115:405–16.

16. Shlyonsky VG, Mies F, Sariban-Sohraby S. Epithelial sodium channel activity in detergent-resistant membrane microdomains. Am J Physiol Renal Physiol. 2003;284:F182–8.

17. Awayda MS, Awayda KL, Pochynyuk O, Bugaj V, Stockand JD, Ortiz RM. Acute cholesterol-induced anti-natriuretic effects: role of epithelial Na⁺ channel activity, protein levels, and processing. J Biol Chem. 2011;286:1683–95.

18. Lockwich TP, Liu X, Singh BB, Jadlowiec J, Weiland S, Ambudkar IS. Assembly of Trp1 in a signaling complex associated with caveolin-scaffolding lipid raft domains. J Biol Chem. 2000;275:11934–42.

19. Chubinskiy-Nadezhdin VI, Negulyaev YA, Morachevskaya EA. Cholesterol depletion-induced inhibition of stretch-activated channels is mediated via actin rearrangement. Biochem Biophys Res Commun. 2011;412:80–5.

20. Bukiya AN, Durdagi S, Noskov S, Rosenhouse-Dantsker A. Cholesterol up-regulates neuronal G protein-gated inwardly rectifying potassium (GIRK) channel activity in the hippocampus. J Biol Chem. 2017;292:6135–47.

21. Yi A, Lin Y-F, Jan YN, Jan LY. Yeast screen for constitutively active mutant G protein–activated potassium channels. Neuron. 2001;29:657–67.

22. Vivaudou M, Chan KW, Sui JL, Jan LY, Reuveny E, Logothetis DE. Probing the G-protein regulation of GIRK1 and GIRK4, the two subunits of the K_ACh channel, using functional homomeric mutants. J Biol Chem. 1997;272:31553–60.

23. Lundbaek JA, Andersen OS. Spring constants for channel-induced lipid bilayer deformations estimates using gramicidin channels. Biophys J. 1999;76:889–95.

24. Lundbaek JA, Birn P, Hansen AJ, Andersen OS. Membrane stiffness and channel function. Biochemistry. 1996;35:3825–30.

25. Rosenhouse-Dantsker A. Insights into the molecular requirements for cholesterol binding to ion channels. Curr Top Membr. 2017;80:187–208.
26. Cherezov V, Rosenbaum DM, Hanson MA, Rasmussen SG, Thian FS, Kobilka TS, Choi HJ, Kuhn P, Weis WI, Kobilka BK, Stevens RC. High-resolution crystal structure of an engineered human beta2-adrenergic G protein-coupled receptor. Science. 2007;318:1258–65.
27. Wada T, Shimono K, Kikukawa T, Hato M, Shinya N, Kim SY, Kimura-Someya T, Shirouzu M, Tamogami J, Miyauchi S, Jung KH, Kamo N, Yokoyama S. Crystal structure of the eukaryotic light-driven proton-pumping rhodopsin, Acetabularia rhodopsin II, from marine alga. J Mol Biol. 2011;411(5):986–98.
28. Liu W, Wacker D, Gati C, Han GW, James D, Wang D, Nelson G, Weierstall U, Katritch V, Barty A, Zatsepin NA, Li D, Messerschmidt M, Boutet S, Williams GJ, Koglin JE, Seibert MM, Wang C, Shah ST, Basu S, Fromme R, Kupitz C, Rendek KN, Grotjohann I, Fromme P, Kirian RA, Beyerlein KR, White TA, Chapman HN, Caffrey M, Spence JC, Stevens RC, Cherezov V. Serial femtosecond crystallography of G protein-coupled receptors. Science. 2013;342:1521–4.
29. Huang W, Manglik A, Venkatakrishnan AJ, Laeremans T, Feinberg EN, Sanborn AL, Kato HE, Livingston KE, Thorsen TS, Kling RC, Granier S, Gmeiner P, Husbands SM, Traynor JR, Weis WI, Steyaert J, Dror RO, Kobilka BK. Structural insights into μ-opioid receptor activation. Nature. 2015;524(7565):315–21.
30. Cheng RKY, Segala E, Robertson N, Deflorian F, Doré AS, Errey JC, Fiez-Vandal C, Marshall FH, Cooke RM. Structures of human A_1 and A_{2A} adenosine receptors with xanthines reveal determinants of selectivity. Structure. 2017;25:1275–85.
31. Che T, Majumdar S, Zaidi SA, Ondachi P, McCorvy JD, Wang S, Mosier PD, Uprety R, Vardy E, Krumm BE, Han GW, Lee MY, Pardon E, Steyaert J, Huang XP, Strachan RT, Tribo AR, Pasternak GW, Carroll FI, Stevens RC, Cherezov V, Katritch V, Wacker D, Roth BL. Structure of the nanobody-stabilized active state of the kappa opioid receptor. Cell. 2018;172:55–67.
32. Penmatsa A, Wang KH, Gouaux E. X-ray structure of dopamine transporter elucidates antidepressant mechanism. Nature. 2013;503:85–90.
33. Coleman JA, Green EM, Gouaux E. X-ray structures and mechanism of the human serotonin transporter. Nature. 2016;532:334–9.
34. Shinoda T, Ogawa H, Cornelius F, Toyoshima C. Crystal structure of the sodium-potassium pump at 2.4 A resolution. Nature. 2009;459:446–50.
35. Marsh D, Barrantes FJ. Immobilized lipid in acetylcholine receptor-rich membranes from Torpedo marmorata. Proc Natl Acad Sci U S A. 1978;75:4329–33.
36. Jones OT, McNamee MG. Annular and nonannular binding sites for cholesterol associated with the nicotinic acetylcholine receptor. Biochemistry. 1988;27:2364–74.
37. Addona GH, Sandermann H Jr, Kloczewiak MA, Husain SS, Miller KW. Where does cholesterol act during activation of the nicotinic acetylcholine receptor? Biochim Biophys Acta. 1998;1370:299–309.
38. Barrantes FJ. Structural basis for lipid modulation of nicotinic acetylcholine receptor function. Brain Res Brain Res Rev. 2004;47:71–95.
39. Levitan I, Singh DK, Rosenhouse-Dantsker A. Cholesterol binding to ion channels. Front Physiol. 2014;5:65.
40. Epand RM. Cholesterol and the interaction of proteins with membrane domains. Prog Lipid Res. 2006;45:279–94.
41. Zhang Z, Tóth B, Szollosi A, Chen J, Csanády L. Structure of a TRPM2 channel in complex with Ca^{2+} explains unique gating regulation. eLife. 2018;7:e36409.
42. Romanenko VG, Rothblat GH, Levitan I. Modulation of endothelial inward rectifier K^+ current by optical isomers of cholesterol. Biophys J. 2002;83:3211–22.
43. Addona GH, Sandermann H Jr, Kloczewiak MA, Miller KW. Low chemical specificity of the nicotinic acetylcholine receptor sterol activation site. Biochim Biophys Acta. 2003;1609:177–82.
44. Singh DK, Rosenhouse-Dantsker A, Nichols CG, Enkvetchakul D, Levitan I. Direct regulation of prokaryotic Kir channel by cholesterol. J Biol Chem. 2009;284:30727–36.

45. D'Avanzo N, Hyrc K, Enkvetchakul D, Covey DF, Nichols CG. Enantioselective protein-sterol interactions mediate regulation of both prokaryotic and eukaryotic inward rectifier K⁺ channels by cholesterol. PLoS One. 2011;6:e19393.

46. Bukiya AN, Belani JD, Rychnovsky S, Dopico AM. Specificity of cholesterol and analogs to modulate BK channels points to direct sterol-channel protein interactions. J Gen Physiol. 2011;137:93–110.

47. Picazo-Juárez G, Romero-Suárez S, Nieto-Posadas A, Llorente I, Jara- Oseguera A, Briggs M, McIntosh TJ, Simon SA, Ladrón-de-Guevara E, Islas LD, Rosenbaum T. Identification of a binding motif in the S5 helix that confers cholesterol sensitivity to the TRPV1 ion channel. J Biol Chem. 2011;286:24966–76.

48. Covey DF. ent-Steroids. Novel tools for studies of signaling pathways. Steroids. 2009;74:577–85.

49. Baier CJ, Fantini J, Barrantes FJ. Disclosure of cholesterol recognition motifs in transmembrane domains of the human nicotinic acetylcholine receptor. Sci Rep. 2011;1:69.

50. Fantini J, Barrantes FJ. How cholesterol interacts with membrane proteins. An exploration of cholesterol-binding sites including CRAC, CARC, and tilted domains. Front Physiol. 2013;4:31.

51. Bukiya AN, Rosenhouse-Dantsker A. Synergistic activation of G protein-gated inwardly rectifying potassium channels by cholesterol and PI(4,5)P₂. Biochim Biophys Acta Biomembr. 2017;1859:1233–41.

52. Rosenhouse-Dantsker A, Noskov S, Durdagi S, Logothetis DE, Levitan I. Identification of novel cholesterol-binding regions in Kir2 channels. J Biol Chem. 2013;288:31154–64.

53. Hibino H, Inanobe A, Furutani K, Murakami S, Findlay I, Kurachi Y. Inwardly rectifying potassium channels: their structure, function, and physiological roles. Physiol Rev. 2010;90:291–366.

54. Reimann F, Ashcroft FM. Inwardly rectifying potassium channels. Curr Opin Cell Biol. 1999;11:503–8.

55. Bichet D, Haass FA, Jan LY. Merging functional studies with structures of inward-rectifier K⁺ channels. Nat Rev Neurosci. 2003;4:957–67.

56. Kubo Y, Adelman JP, Clapham DE, Jan LY, Karschin A, Kurachi Y, Lazdunski M, Nichols CG, Seino S, Vandenberg CA. International Union of Pharmacology. LIV. Nomenclature and molecular relationships of inwardly rectifying potassium channels. Pharmacol Rev. 2005;57:509–26.

57. Nichols C, Lopatin A. Inward rectifier potassium channels. Annu Rev Physiol. 1997;59:268–77.

58. Ho K, Nichols CG, Lederer WJ, Lytton J, Vassilev PM, Kanazirska MV, et al. Cloning and expression of an inwardly rectifying ATP-regulated potassium channel. Nature. 1993;362:31–8.

59. Dascal N, Schreibmayer W, Lim NF, Wang W, Chavkin C, DiMagno L, et al. Atrial G protein-activated K⁺ channel: Expression cloning and molecular properties. Proc Natl Acad Sci U S A. 1993;90:10235–9.

60. Kubo Y, Baldwin TJ, Jan YN, Jan LY. Primary structure and functional expression of a mouse inward rectifier potassium channel. Nature. 1993;362:127–33.

61. Nishida M, Cadene M, Chait BT, MacKinnon R. Crystal structure of a Kir3.1- prokaryotic Kir channel chimera. EMBO J. 2007;26:4005–15.

62. Tao X, Avalos JL, Chen J, MacKinnon R. Crystal structure of the eukaryotic strong inward-rectifier K⁺ channel Kir2.2 at 3.1 A resolution. Science. 2009;326:1668–74.

63. Whorton MR, MacKinnon R. Crystal structure of the mammalian GIRK2 K⁺ channel and gating regulation by G proteins, PIP₂, and sodium. Cell. 2011;147:199–208.

64. Li N, Wu JX, Ding D, Cheng J, Gao N, Chen L. Structure of a Pancreatic ATP-Sensitive Potassium Channel. Cell. 2017;168:101–10.

65. Lee KPK, Chen J, MacKinnon R. Molecular structure of human KATP in complex with ATP and ADP. Elife. 2017;6:e32481.

66. Tikku S, Epshtein Y, Collins H, Travis AJ, Rothblat GH, Levitan I. Relationship between Kir2.1/Kir2.3 activity and their distributions between cholesterol-rich and cholesterol-poor membrane domains. Am J Physiol Cell Physiol. 2007;293:C440–50.

67. Chen X, Johnston D. Constitutively active G-protein-gated inwardly rectifying K⁺ channels in dendrites of hippocampal CA1 pyramidal neurons. J Neurosci. 2005;25:3787–92.
68. VanDongen AM, Codina J, Olate J, Mattera R, Joho R, Birnbaumer L, Brown AM. Newly identified brain potassium channels gated by the guanine nucleotide binding protein G₀. Science. 1988;242:1433–7.
69. Leaney JL. Contribution of Kir3.1, Kir3.2A and Kir3.2C subunits to native G protein-gated inwardly rectifying potassium currents in cultured hippocampal neurons. Eur J Neurosci. 2003;18:2110–8.
70. Slesinger PA, Patil N, Liao YJ, Jan YN, Jan LY, Cox DR. Functional effects of the mouse weaver mutation on G protein-gated inwardly rectifying K⁺ channels. Neuron. 1996;16:321–31.
71. Inanobe A, Yoshimoto Y, Horio Y, Morishige KI, Hibino H, Matsumoto S, Tokunaga Y, Maeda T, Hata Y, Takai Y, Kurachi Y. Characterization of G-protein-gated K⁺ channels composed of Kir3.2 subunits in dopaminergic neurons of the substantia nigra. J Neurosci. 1999;19:1006–17.
72. Rubinstein M, Peleg S, Berlin S, Brass D, Keren-Raifman T, Dessauer CW, Ivanina T, Dascal N. Divergent regulation of GIRK1 and GIRK2 subunits of the neuronal G protein gated K⁺ channel by GalphaiGDP and Gbetagamma. J Physiol. 2009;587(Pt 14):3473–91.
73. Logothetis DE, Jin T, Lupyan D, Rosenhouse-Dantsker A. Phosphoinositide-mediated gating of inwardly rectifying K(+) channels. Pflugers Arch. 2007;455:83–95.
74. Hansen SB, Tao X, MacKinnon R. Structural basis of PIP2 activation of the classical inward rectifier K+ channel Kir2.2. Nature. 2011;477:495–8.
75. Schmidt MR, Stansfeld PJ, Tucker SJ, Sansom MS. Simulation-based prediction of phosphatidylinositol 4,5-bisphosphate binding to an ion channel. Biochemistry. 2013;52:279–81.
76. Yeagle PL. Non-covalent binding of membrane lipids to membrane proteins. Biochim Biophys Acta. 2014;1838:1548–59.
77. Hanson MA, Cherezov V, Griffith MT, Roth CB, Jaakola VP, Chien EYT, Velasquez J, Kuhn P, Stevens RC. A specific cholesterol binding site is established by the 2.8Å structure of the human β-adrenergic receptor. Structure. 2008;16:897–905.
78. Manglik A, Kruse AC, Kobilka TS, Thian FS, Mathiesen JM, Sunahara RK, Pardo L, Weis WI, Kobilka BK, Granier S. Crystal structure of the micro-opioid receptor bound to a morphinan antagonist. Nature. 2012;485:321–6.
79. Liu W, Chun E, Thompson AA, Chubukov P, Xu F, Katritch V, Han GW, Roth CB, Heitman LH, Jzerman AP, Cherezov V, Stevens RC. Structural basis for allosteric regulation of GPCRs by sodium ions. Science. 2012;337:232–6.
80. Wacker D, Wang C, Katritch V, Han GW, Huang XP, Vardy E, McCorvy JD, Jiang Y, Chu M, Siu FY, Liu W, Xu HE, Cherezov V, Roth BL, Stevens RC. Structural features for functional selectivity at serotonin receptors. Science. 2013;340:615–9.
81. Li H, Papadopoulos V. Peripheral-type benzodiazepine receptor function in cholesterol transport. Identification of a putative cholesterol recognition/interaction amino acid sequence and consensus pattern. Endocrinology. 1998;139:4991–7.
82. Singh AK, McMillan J, Bukiya AN, Burton B, Parrill AL, Dopico AM. Multiple cholesterol recognition/interaction amino acid consensus (CRAC) motifs in cytosolic C tail of Slo1 subunit determine cholesterol sensitivity of Ca²⁺- and voltage-gated K⁺ (BK) channels. J Biol Chem. 2012;287:20509–21.
83. Epshtein Y, Chopra AP, Rosenhouse-Dantsker A, Kowalsky GB, Logothetis DE, Levitan I. Identification of cholesterol sensitive domain in the C-terminus of Kir2.1 channels. Proc Natl Acad Sci U S A. 2009;106:8055–60.
84. Jiang Y, Lee A, Chen J, Cadene M, Chait BT, MacKinnon R. The open pore conformation of potassium channels. Nature. 2002;417:523–6.
85. Jin T, Peng L, Mirshahi T, Rohacs T, Chan KW, Sanchez R, Logothetis DE. The βγ subunits of G proteins gate a K⁺ channel by pivoted bending of a transmembrane segment. Mol Cell. 2002;10:469–81.
86. Rosenhouse-Dantsker A, Logothetis DE. New roles for a key glycine and its neighboring residue in potassium channel gating. Biophys J. 2006;91:2860–73.
87. Rosenhouse-Dantsker A. Cholesterol-binding sites in GIRK channels: the devil is in the details. Lipid Insights. 2018;11:1178635317754071.

Insights into the Molecular Mechanisms of Cholesterol Binding to the NPC1 and NPC2 Proteins

Stephanie M. Cologna and Avia Rosenhouse-Dantsker

Abstract In recent years, a growing number of studies have implicated the coordinated action of NPC1 and NPC2 in intralysosomal transport and efflux of cholesterol. Our current understanding of this process developed with just over two decades of research. Since the cloning of the genes encoding the NPC1 and NPC2 proteins, studies of the biochemical defects observed when either gene is mutated along with computational and structural studies have unraveled key steps in the underlying mechanism. Here, we summarize the major contributions to our understanding of the proposed cholesterol transport controlled by NPC1 and NPC2, and briefly discuss recent findings of cholesterol binding and transport proteins beyond NPC1 and NPC2. We conclude with key questions and major challenges for future research on cholesterol transport by the NPC1 and NPC2 proteins.

Keywords Sterol-sensing domain · Transport · Binding · Structure · Lysosome · Niemann-Pick Disease Type C

Abbreviations

CLR   Cholesterol
CTD   C-terminal domain
MLD   Middle luminal domain

S. M. Cologna (✉)
Department of Chemistry, University of Illinois at Chicago, Chicago, IL, USA

Laboratory for Integrative Neuroscience, University of Illinois at Chicago, Chicago, IL, USA
e-mail: cologna@uic.edu

A. Rosenhouse-Dantsker (✉)
Department of Chemistry, University of Illinois at Chicago, Chicago, IL, USA
e-mail: dantsker@uic.edu

© Springer Nature Switzerland AG 2019
A. Rosenhouse-Dantsker, A. N. Bukiya (eds.), *Direct Mechanisms in Cholesterol Modulation of Protein Function*, Advances in Experimental Medicine and Biology 1135, https://doi.org/10.1007/978-3-030-14265-0_8

NPC   Niemann-Pick Type C
NTD   N-terminal domain
SSD   Sterol sensing domain

1   Overview

Cholesterol (Fig. 1a) plays a critical role in multiple cellular functions. These include regulating the physical properties of the plasma membrane to ensure cell viability, growth, proliferation, and serving as a signaling and precursor molecule in biochemical pathways [1–7]. Regulation of cellular cholesterol levels is tightly controlled via multiple pathways that include *de-novo* biosynthesis, uptake, recycling and release [7–11]. In particular, delivery of extracellular cholesterol to cells is achieved by receptor-mediated uptake of low density lipoproteins (LDLs) that carry both free and esterified cholesterol [12]. Subsequent to entering the vascular tissue, LDL particles reach the endosomal-lysosomal system via endocytosis [13]. In this system, cholesterol esters are converted back to free cholesterol. Proteins are then harnessed to export the hydrophobic cholesterol molecule through the hydrophilic environment of endosomes and lysosomes. In particular, coordination of the NPC1 and NPC2 proteins (Fig. 2) facilitates cholesterol trafficking through the lysosome. Genetic mutations of either gene result in the accumulation of unesterified cholesterol in the endo-lysosomal system [14]. While clinical manifestations were reported in the early 1900's [15, 16], it was not until the 1980s that an understanding of the relationship between the genetic defect and the clinical phenotype began to emerge [17–28]. This chapter introduces the disease component related to the proteins involved in lysosomal cholesterol trafficking as well as covers the cloning and structural discoveries that represent our current understanding of cholesterol movement via NPC1 and NPC2 interplay.

2   Discovery of NPC Disease

The discovery of Niemann-Pick Type C (NPC) disease occurred in the early 1900s by Albert Niemann reporting combined hepatosplenomegaly and central nervous system defects in a young child [15] (also reviewed in [29]). Clinical and pathological evaluation continued in the 1920s when the distinct disease was established by Ludwig Pick [16]. In 1961, Crocker proposed to classify Niemann-Pick Disease into Types A, B, C and D based upon differential clinical and biochemical phenotypes [30]. Further studies have linked Types A and B to mutations in the *SMPD1* gene and sphingomyelinase deficiency, and are now regarded as a distinct disorder [31]. Collectively, today NPC represents the previously termed C and D Types and includes progressive cerebellar neurodegeneration, which manifests in ataxia, seizures, enlarged liver and spleen and other clinical features (as reviewed in [32]). To

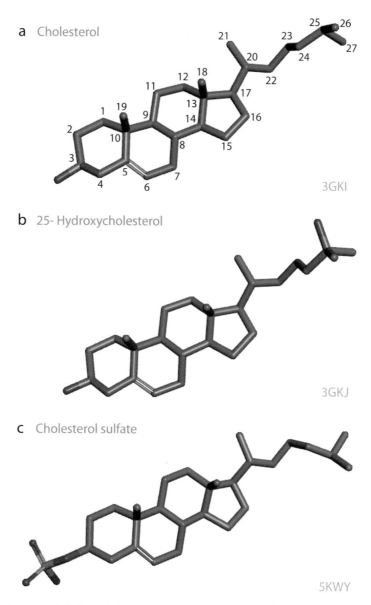

Fig. 1 Structure of cholesterol, 25-hydroxycholesterol and cholesterol sulfate. Stick representation of (**a**) cholesterol based on the cholesterol molecule bound to the N-terminal domain of NPC1 (PDB ID: 3GKI) showing the numbering of the carbon atoms in the molecule, (**b**) 25-hydroxycholesterol based on the 25-hydroxycholesterol molecule bound to the N-terminal domain of NPC1 (PDB ID: 3GKJ), and (**c**) cholesterol sulfate based on the cholesterol sulfate molecule bound to NPC2 (PDB ID: 5KWY)

Fig. 2 Structure of NPC1
and NPC2. (**a**) Cartoon
representation of NPC1 as
constructed by aligning the
2017 Li et al. [81] crystal
structure (PDB ID 5U73)
and the 2016 Gong et al.
[73] cryo EM structure
(PDB ID 3JD8). The
structure depicted includes
TM2-13, the middle
luminal domain, and the
C-terminal domain based
on PDB ID 5U73. TM1
and the N-terminal domain
are based on PDB ID
3JD8. Critical domains for
cholesterol binding are
shown including the sterol
sensing domain (SSD)
comprised of TM3-7 (in
light cyan), the N-terminal
domain (NTD) (in pink),
the middle luminal domain
(MLD) (in green), and the
C-terminal domain (CTD)
(in yellow). (**b**) Cartoon
representation of NPC2
based on the 2007 Xu et al.
structure (PDB ID 2HKA)

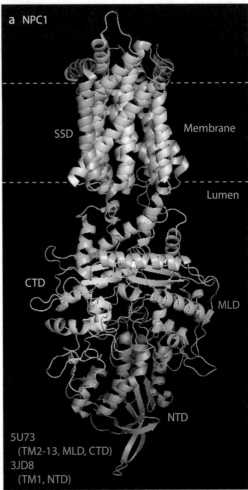

indicate the genetic cause of the disease, the field has more recently taken forth the notations of NPC1 and NPC2 to represent the genetic causation of NPC disease.

Pioneering work by Roscoe Brady and Peter Pentchev [33, 34] provided biochemical insights into the function of both NPC1 and NPC2. In the spontaneous occurring NPC mouse model, cholesterol storage, lysosomal enzyme activity defects and similarities with the human described disease were observed [18, 35]. Moreover, esterification of cholesterol was defective in this mouse model similarly to the case in NPC patients [17]. Other biochemical features of patient cultured fibroblasts and in the NPC mouse model included alterations of LDL-mediated processes [19, 36, 37], demyelination [38, 39], cerebellar degeneration [40–43], oxidative stress [44, 45], altered calcium homeostasis [46–48] and others [49–52]. Additionally, further studies have demonstrated that many cell types are defective in NPC [53, 54]. Notably, NPC1 and NPC2 mutations lead to similar NPC disease phenotypes [55] suggesting that the two proteins may function either together or sequentially in a common pathway affecting cholesterol transport.

3   NPC2 Gene and Protein Structure

The NPC2 protein, also known as HE1, was reported in 2000 by Peter Lobel's group to be the other gene affected in NPC [28]. NPC2 is a small, ubiquitous, lysosomal protein that is often found in epididymis fluid and was cloned in the late 1990's [56, 57]. In the latter study, the NPC2 protein was shown to bind cholesterol in a porcine model with a 1:1 stoichiometry and micromolar affinity [57]. The crystal structure of the protein at 1.7 Å was reported in 2003 revealing an immunoglobulin-like β-sandwich fold consisting of seven β-strands arranged in two β-sheets forming a loosely packed hydrophobic core (Fig. 2b) [58]. This observation led to the suggestion that the hydrophobic core constitutes an incipient internal cholesterol binding pocket [58]. However, as potential hydrophobic pockets in NPC2 were too small to accommodate cholesterol, it was subsequently proposed that a shift in the two β sheets of NPC2 would be necessary upon cholesterol binding [59]. Specific clues to the location of the putative binding pocket of cholesterol in NPC2 were obtained from mutagenesis studies showing that the F66A, V96F, and Y100A mutations in NPC2 lead to a decrease in cholesterol binding to the protein *in vitro*. When added to NPC2 deficient cells, these mutants were unable to clear elevated cholesterol levels, further supporting the notion that the ability of NPC2 to bind cholesterol is necessary for normal protein function. However, with the discovery that the K32A, D72A, and K75A NPC2 mutants all exhibited normal cholesterol binding, but were unable to correct the cholesterol accumulation phenotype of the cells, it was proposed that cholesterol binding may not be the only requirement for normal NPC2 function [59].

An indication that NPC2 may have a cholesterol transport function emanated from experiments demonstrating that the absolute cholesterol transfer rates from NPC2 to the membrane were orders of magnitude faster than its off-rates from

NPC2 to an aqueous buffer [59]. These experiments also led to the hypothesis that subsequent to binding cholesterol from internal lysosomal membranes, NPC2 interacts with NPC1, thereby facilitating post-lysosomal export of cholesterol [59]. The idea that NPC2 alone was not sufficient for cholesterol egress from lysosomes was further supported by genetic considerations [33, 60]. Additional experiments demonstrated that the rate of transfer of cholesterol from NPC2 to membrane vesicles increased with the frequency of NPC2-membrane electrostatic interactions, particularly in an acidic environment such as that in lysosomes, supporting the proposed role of NPC2 in lysosomal cholesterol transport [61]. Consequently, NPC2 could significantly accelerate the rates of cholesterol transport from and between membranes, as well as the extent of cholesterol transfer. It was shown that transfer of cholesterol occurred rapidly via direct NPC2-membrane interactions via a collisional mechanism, and suggested that NPC2 bound to the membrane surface without penetration into the bilayer hydrophobic core [61]. More recently, in a study carried out in 2015 by the Storch laboratory, it was demonstrated that multiple different mutations in several surface regions of NPC2 exhibited deficient cholesterol transport properties, and were unable to promote egress of accumulated intracellular cholesterol from NPC2 knock out fibroblasts [62]. The point mutations caused changes in the surface charge distribution of NPC2 with minimal conformational changes. Furthermore, complementary molecular modeling showed that NPC2 was highly plastic, with several positively charged regions across the surface that could interact with negatively charged membrane phospholipids. This led the authors to suggest that the plasticity of NPC2 may allow for multiple mechanisms for sterol transfer, and that NPC2 could bind to more than one membrane simultaneously. Consequently, NPC2 may act to traffic cholesterol rapidly at zones of close apposition between membranes such as those that exist in the interior of endo/lysosomes [62].

Structural insights into the molecular basis for sterol binding by NPC2 were obtained in 2007 when NPC2 was co-crystallized with cholesterol sulfate (Fig. 1c) at a resolution of 1.81 Å [63] (Table 1). The sulfate moiety was the only portion of the ligand exposed to solvent, peeking out of the hydrophobic sterol binding pocket.

Table 1 Structures of NPC1 and/or NPC2 in complex with cholesterol or cholesterol derivatives

PDB ID	Protein	Ligand	Release date	Resolution (Å)
2HKA	NPC2	C3S (cholesterol sulfate)	6/26/2007	1.81
3GKI	NPC1 N-terminal domain (NTD)	CLR (cholesterol)	7/14/2009	1.8
3GKJ	NPC1 N-terminal domain (NTD)	HC3 (25-hydroxycholesterol)	7/14/2009	1.6
3JD8	NPC1	CLR (cholesterol)	6/1/2016	4.43
5KWY	NPC1 middle lumenal domain (MLD) bound to NPC2	C3S (cholesterol sulfate)	8/24/2016	2.4

In contrast, the cholesterol iso-octyl tail was shielded from the hydrophilic environment by the interior residues of the NPC2 protein. Comparison between the holo NPC2 structure with a bound cholesterol sulfate and the apo NPC2 structure corroborated that NPC2 had a sterol incipient binding pocket, which was formed from several adjacent small cavities that expanded to accommodate the closely sequestered steroid nucleus of cholesterol sulfate. The structures showed that during this process, the β-strands of NPC2 separated slightly while undergoing substantial side chain reorientation [63].

4   NPC1 Gene Cloning, Structure and Cholesterol Binding

While the pathology of NPC disease as a sterol transport defect was described already in the early 1900s [15, 16], the genetic cause remained unknown until the cloning of the *NPC1* gene in both humans and mice in the late 1990s [27, 64]. Since these reports, and with the implementation of gene sequencing in clinical research, it has become evident that the majority of NPC patients have NPC1 mutations [65]. However, while the structural and functional basis for the consequences of genetic mutations in the NPC2 protein began to unravel in the late 1990's, the structure and role of the NPC1 protein in cholesterol binding and transport remained unknown for several more years.

Initial clues regarding the role of NPC1 started to emerge in 1999 in a mutagenesis study by Watari et al. that suggested that the transmembrane region encompassing helices 3-7, which is thought to form a sterol sensing domain (SSD) (Fig. 2), was required for normal cholesterol egress from the endosome/lysosome system [66]. This concept was further supported by a 2004 study by Ohgami et al., in which a photoactivatable cholesterol analog was implicated in binding to NPC1 with low affinity [67]. The study demonstrated that the SSD was required for NPC1 to bind the cholesterol analog. It was thus suggested that NPC1 may be involved in cholesterol transport at the late endosomal membrane and/or that cholesterol may regulate the activity of NPC1. However, whether and how NPC1 function was linked to NPC2 function remained unclear. While it was proposed in 2001 that NPC1 activity may depend on NPC2 [68], it was found that the interaction between the photoactivatable cholesterol analog and NPC1 did not require NPC2 [67].

In a subsequent study carried out by the Brown and Goldstein labs in 2008, the N-terminal domain (NTD) of NPC1 (Fig. 2a) was implicated as a sterol binding site with a sub-micromolar affinity [69]. Comparison of the differential ability of a variety of oxysterols to bind to NPC1 suggested that in contrast to the orientation of cholesterol when it binds to NPC2, upon binding to NPC1, the hydroxyl group of the cholesterol molecule faces the interior of the NTD of NPC1 whereas the iso-octyl tail is exposed [69]. In 2009, the cholesterol-bound structure of the NTD of NPC1 was solved using X-ray crystallography at a resolution of 1.8 Å confirming the predicted orientation of cholesterol in the NTD binding site [70] (Table 1). As a comparison, the structures of the apo and the 25-hydroxycholesterol (Fig. 1b) bound NTD of NPC1

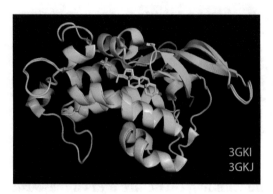

Fig. 3 Cholesterol and 25-hydroxycholesterol share the same binding site in the N-terminal domain of NPC1. Alignment of the structures of the cholesterol- and 25-hydroxycholesterol-bound N-terminal domain of NPC1 in the 2009 Kwon et al. [70] structures (PDB IDs 3GKI and 3GKJ, respectively). The protein is depicted in ribbon representation, and the sterols in stick representation

were also determined. Cholesterol and 25-hydroxycholesterol bind in a similar manner to the NTD of NPC1, interacting with the same protein residues (Fig. 3). The 25-hydroxyl group on the 25-hydroxycholesterol molecule formed a water-mediated interaction with the main chain of L175. The sterol binding pocket was lined primarily with hydrophobic residues including W27, L83, F108, P202, F203, and I205. Two polar residues, N41 and Q79 formed hydrogen bonding with the cholesterol hydroxyl group. E30 formed a water-mediated interaction with the hydroxyl group of the cholesterol molecule, thereby stabilizing the interaction between the NTD of NPC1 and cholesterol, and imposing stereospecificity. When added to NPC1-deficient cells, alanine mutants of these residues failed to restore function. At each end of the sterol-binding pocket was an opening toward the surrounding solvent. One opening was located near the cholesterol hydroxyl group, and was large enough for a single water molecule to enter or exit. In contrast, the second opening was located at the end of the cholesterol iso-octyl side chain, and was not large enough to permit passage of the tetracyclic ring without a conformational change indicating that it would need to expand to facilitate cholesterol entry [70].

The growing structural and functional insights into the binding of cholesterol to the NTD of the NPC1 protein, along with the realization that the orientation of cholesterol binding to this NPC1 domain was opposite to the way that cholesterol binds to NPC2, supported the notion that there is cooperation among NPC1 and NPC2 in cholesterol transport. Accordingly, reversal of the orientation of cholesterol during its transfer from NPC2 to NPC1 would allow its iso-octyl hydrophobic tail to lead the way into the outer lysosomal membrane [70].

In a parallel line of research, in vitro work raised the possibility that the NTD may not be the only cholesterol binding site in NPC1, and that NPC1 could possess a second binding site, possibly at the SSD [71]. NPC1 with alanine point mutations of L175/L176, D180/D182, N185, T187/N188, E191/Y192, and G199/Q200 in a helical subdomain of the SSD consisting of helices 7, 8 and the intervening loop could not restore cholesterol exit from lysosomes in NPC1-deficient cells. In line with these results, a later study that explored the binding of oxysterol derivatives to NPC1 demonstrated that they bound directly and selectively to a low-affinity or transient non-NTD sterol binding site [72]. In 2016, two structures that included the SSD domains were utilized to further explore the possibility of a second sterol binding site in the SSD [73, 74]. The first was a cryo-EM structure of the full length NPC1 that was obtained at 4.4 Å resolution [73] (Table 1). The structure showed that the five SSD-forming membrane helices (3–7) were exposed to the lipid bilayer suggesting that the SSD was available for potential interactions with membrane-embedded sterols. The second structure was a crystallographic structure obtained at 3.6 Å resolution, and included 12 of the 13 transmembrane domains of NPC1 [74]. In this structure, the NPC1 SSD formed a cavity that was accessible from both the luminal bilayer leaflet and the endosomal lumen. Complementary computational modeling suggested that this cavity was large enough to accommodate one cholesterol molecule, further supporting the notion that the SSD may harbor a second cholesterol binding site [74]. Combining the accumulating evidence that NPC2, the NTD of NPC1, and the SSD of NPC1 all possessed a sterol binding site paved the way to a comprehensive model of the molecular mechanism of cholesterol transport by NPC1 and NPC2 in the lysosomes involving cholesterol derived LDL uptake in the lysosome, followed by binding to NPC2 which would then 'handoff' cholesterol to NPC1 for recycling out of the lysosome.

5 Development of a Cholesterol Transport Mechanism Model from NPC2 to NPC1

In 2008, the Brown and Goldstein laboratories demonstrated that cholesterol can transfer between the NTD of NPC1 and NPC2 in a bidirectional fashion facilitated by NPC2 [75]. While the transfer of cholesterol to and from NPC2 was rapid, the transfer of cholesterol to and from the NTD of NPC1 was very slow. The latter, however, was significantly accelerated in the presence of NPC2, supporting the notion that the two proteins act together to facilitate cholesterol egress from lysosomes [75].

Then, subsequent to determining the structure of the cholesterol-bound NTD of NPC1 in 2009 [70], the groups of Brown and Goldstein further evaluated in 2010 the cholesterol 'handoff' mechanism between NPC2 and the NTD of NPC1, and showed that in the presence of mutations of surface residues such as V81 of NPC2 and L175/L176 of the NTD of NPC1, cholesterol binding could occur but 'handoff' did not [76]. This raised the possibility that these 'transfer mutants' that clustered in surface patches of NPC2 and the NTD of NPC1 interacted with each other to facilitate the opening of the binding pocket in the NTD of NPC1, thereby allowing cholesterol to transfer between the two proteins. Further support of the notion that NPC2 and NPC1 interacted with each other came in 2011 from the Pfeffer laboratory who showed that NPC2 directly interacted with the NPC1 middle luminal domain (MLD) (Fig. 2a) in an acidic environment of pH 5.5 with a low micromolar affinity and a cholesterol dependent binding strength that increased when NPC2 was carrying a cholesterol molecule [77]. In alignment with these observations, the disease causing mutations R404Q and R518Q in the NPC1 MLD interfered with the ability of NPC1 to bind to NPC2. These results stimulated the idea that the NPC1 MLD may bring NPC2 into close proximity with the NPC1 NTD to facilitate the transfer of cholesterol between NPC2 and the NTD of NPC1. Once transferred, the loss of cholesterol from NPC2 would trigger its release from the MLD following a reduction in the binding strength between the two proteins. The NPC1 NTD-bound cholesterol would then be transferred to the lysosomal membrane bilayer [77].

In 2013, the Wiest group carried out computational studies of the NPC2-NTD(NPC1) system to gain further mechanistic insights into cholesterol binding and transfer between the two proteins (Fig. 4) [78] based on working models from the Brown and Goldstein laboratories [75]. The results of the simulations suggested that when bound to NPC2, the cholesterol hydroxyl group formed multiple interactions with NPC1 residues, thereby stabilizing the interaction between the NTD of NPC1 and NPC2. The simulations also suggested that a large reorganization occurred in the binding pocket of the NPC1 NTD upon cholesterol binding. These results inspired the 'sliding model' that enhanced the mechanistic description of the 'handoff model'. According to this model, cholesterol binding to NPC2 would lead to an increase in the association constant for the formation of the complex between NPC2 and the NTD of NPC1. Once the NPC1-NPC2 complex was formed, cholesterol would transfer from NPC2 to the NTD of NPC1. This transfer would proceed through the displacement of multiple NTD helices, followed by the actual transfer of cholesterol through the opened pathway. The transfer of the cholesterol molecule from the NPC2 protein would then lead to a decrease in the association constant of

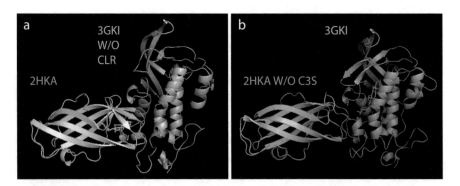

Fig. 4 Models of the putative interactions between the N-terminal domain of NPC1 and NPC2, and the cholesterol binding sites in each protein. (**a**) Ribbon representation of the final structure of the 2013 Wiest et al. [78] simulation of the docked structures of NPC2 (PDB ID 2HKA) and the N-terminal domain of NPC1 (PDB ID 3GKI) with a stick representation of cholesterol bound to NPC2 based on a 2008 working model by Infante et al [75]. Based on bi4005478_si_005.pdb from the Supplementary information of [78]. (**b**) Ribbon representation of the final structure of the 2013 Wiest et al. [78] simulation of the docked structures of NPC2 (PDB ID 2HKA) and the N-terminal domain of NPC1 (PDB ID 3GKI) with a stick representation of cholesterol bound to the N-terminal domain of NPC1 based on a 2008 working model by Infante et al [74]. Based on bi4005478_si_003.pdb from the Supplementary information of [78]

the protein-protein complex, facilitating the dissociation of the complex, and the transport of the cholesterol molecule by NPC1 through the membrane [78].

A follow-up simulation study carried out by Elghobashi-Meinhardt in 2014 explored potential isomerization of the C17–C20–C22–C23 dihedral angle in the tail of the cholesterol molecule (Fig. 1a) during its transfer from NPC2 to the NTD of NPC1 [79]. The cholesterol molecule sampled different geometries inside the binding pockets of the NPC1–NPC2 complex during the simulations [78, 79]. In the final structures of the simulations carried out by the Wiest group [78], the C17–C20–C22–C23 dihedral angle was 71.6° for cholesterol in NPC2 but −157.3° in the NTD(NPC1) binding pocket. Notably, in the respective crystal structures of NPC2 and NTD(NPC1) in complex with cholesterol or a cholesterol derivative, the value of the C17–C20–C22–C23 dihedral angle was nearly identical ranging from −162.2° to −164.5° (NPC2 in complex with cholesterol sulfate (PDB ID 2HKA): −164.5°; NPC1 NTD in complex with cholesterol or 25-hydroxycholesterol (PDB IDs 3GKI, 3GKJ): −163.9° and −162.2°, respectively). Further simulations suggested that cholesterol may isomerize in the NPC2 pocket either before or after docking to the NTD of NPC1 to ensure an efficient transfer. By calculating the energy barrier for rotation of the C17–C20–C22–C23 dihedral angle during the sliding of cholesterol from NPC2 to NTD(NPC1), the likely 'reaction pathway' for cholesterol transfer was predicted. The energy barrier along that path was ~22 kcal/mol in total [79]. The primary contribution to the energy barrier was attributed to the distorted geometry of the tail of the cholesterol molecule within the constrained binding pocket in the NTD of NPC1. This energy barrier was in agreement with semi quantitative experimental kinetic rates corresponding to half-lives of up to ~100 sec at 37 °C [79]. Further studies are required to corroborate experimentally the possibility that cholesterol undergoes isomerization of its tail to facilitate its transfer between NPC2 and NPC1.

In addition to providing insights into the structure of the SSD of NPC1, the 2016 cryo-EM structure of the full length NPC1 determined by Gong et al. and discussed above provided insights into the mechanism of cholesterol transfer from NPC2 to NPC1 delineating the structural relationship between the NTD and the MLD of NPC1 [73]. In this structure, a number of polar and charged residues in the NTD (e.g. Q88, Q92, R96) and MLD (e.g. R518) appeared to form an interface between these two luminal domains. Notably, the single point disease mutations R518W or R518Q in the MLD led to reduced interaction between an isolated MLD construct and NPC2. Similarly, the NPC1 mutants L175A/L176A and Q88A/Q92A/R96A showed decreased binding to NPC2 at pH 6.0. While interaction between NPC2 and an isolated NPC1 NTD was not detected, deletion of the NTD resulted in reduced binding between NPC1 and NPC2. The proximity between L175/L176 and the interface residues suggested that the NTD and the MLD of NPC1 may together constitute a docking site for orienting NPC2, thereby facilitating the transfer of cholesterol to the pocket of NTD [73].

This role of the involvement of the MLD of NPC1 was further supported by a crystal structure of the complex of the MLD of NPC1 and NPC2 with a bound cholesterol sulfate molecule that was determined by the Pfeffer laboratory at 2.4 Å resolu-

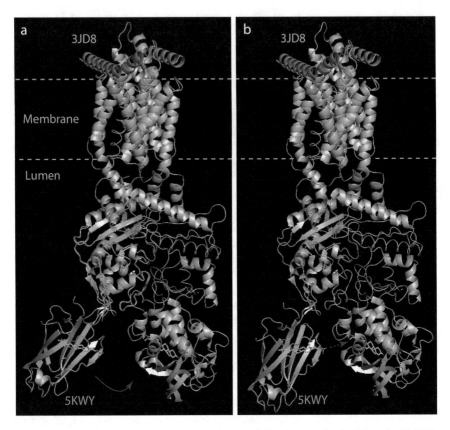

Fig. 5 Model of the putative cholesterol transfer from NPC2 to the N-terminal domain of NPC1. Ribbon representations of the NPC1-NPC2 complex obtained by aligning the structure of the middle luminal domain of NPC1 from the 2016 Li et al. [80] MLD(NPC1)-NPC2 complex (PDB ID 5KWY) with the 2016 Gong et al. [73] cryo-EM structure (PDB ID 3JD8) of NPC1. The MLD(NPC1)-NPC2 complex is in violet, and the NPC1 cryo-EM structure is in gray. The sterol molecules are shown in stick representation (cyan). The alignment in (**a**) is based on the middle luminal domain, and the alignment in (**b**) is based on the two loops of the middle luminal domain that interact with NPC2. By aligning the structures based on these loops, NPC2 approaches closer to the NPC1 N-terminal domain to form a sterol transfer tunnel leading from NPC2 to NPC1. For clarity, the middle luminal domain from the MLD(NPC1)-NPC2 complex was removed in (**b**)

tion in the same year [80] (Table 1). The complexed structure revealed that the MLD of NPC1 binds the top of the NPC2 sterol-binding pocket. Aligning the MLD(NPC1)-NPC2 complex onto the full-length NPC1 cryo-EM structure by aligning the MLDs in the two structures suggested a spatial proximity between NPC2 and the NTD of NPC1 (Fig. 5). The distance between NPC2 and the NTD of NPC1 was further decreased and formed a cholesterol-transfer tunnel when the alignment was done based on the two protruding loops of the MLD of NPC1 that comprised its principal binding interface with NPC2 (Fig. 5) [80]. This interface involved interactions between both polar and hydrophobic residues. For example, interactions were formed between Q421 in the MLD of NPC1 and Q146 of NPC2, and between Y423 of the MLD of NPC1 and M79 of NPC2. Hydrophobic interactions involving F503 and

F504 of the MLD of NPC1 and NPC2 were also observed. Alanine mutations of the equivalent two residues in a murine model significantly reduced the binding affinity between NPC1 MLD and NPC2. Further reduction in binding affinity was observed following mutation of polar MLD residues (e.g. the equivalent residue to Q421) in addition to the above aromatic ones. Notably, a significant rearrangement of the side-chains of NPC2 residues (e.g. K25, M79, K123, and Q146) that contributed substantially to the interface between the MLD of NPC1 and NPC2 was observed in the cholesterol sulfate bound NPC2 structure compared to the apo structure of NPC2. The orientations of the sterol molecules in the binding pockets in NPC2 and in the NTD of NPC1 were compatible with molecular transfer supporting the 'hydrophobic handoff' transfer model between NPC2 and NPC1, and led to the proposal that NPC2 binding to NPC1 may trigger a conformational change(s) that would reorient the NTD of NPC1 into a more planar configuration in relation to NPC2 to accomplish actual cholesterol transfer, as previously modeled [76]. It is important to note, however, that the orientation of the NTD of NPC1 relative to NPC2 in this model [80] somewhat differed from prior computational models (Fig. 4) [78], indicating that further studies are required to establish the precise manner by which NPC2 interacts with the NTD of NPC1 to facilitate cholesterol transfer between the two proteins. With that in mind, the picture emerging from this study was that subsequent to binding cholesterol, NPC2 underwent a subtle conformational change that enhanced its binding to the MLD of NPC1, which in turn led to interactions of NPC2 with the NTD of NPC1. These interactions then slightly impacted the structure and orientation of the MLD protruding loops to orient the NPC2 pocket directly adjacent to the cholesterol binding pocket in the NTD of NPC1 forming a cholesterol transfer tunnel. After the transfer of the cholesterol molecule, the prominent NPC2 residues located at the interface with the MLD of NPC1 would revert to their apo NPC2 conformation, thereby triggering the release of NPC2.

Further insights into the emerging mechanism of lysosomal cholesterol transport was provided by the crystal structure of NPC1 released by the Blobel laboratory, also in 2016 (see above) [74]. This structure included 12 of the 13 transmembrane domains of the protein and extended the proposed model for NPC1 function in cholesterol sensing and transport to include the SSD. According to this extended model, cholesterol would bind first to NPC2, which would then dock to the NTD of NPC1, permitting cholesterol transfer between their binding pockets. The link between the NTD and the remainder of NPC1 would permit the NTD to reorient in such a way that would allow cholesterol transfer to the transmembrane domain. Luminal entry would then be used to transfer cholesterol from its binding site in the NTD of NPC1 that could then exit via the membrane pocket's lateral opening. This lateral opening in the membrane could also provide access for free cholesterol from the lipid bilayer (Fig. 6). It was further proposed that the access properties of the putative binding pocket in the SSD of NPC1 may enable monitoring of the concentration of cholesterol in the lysosome lipid bilayer [74]. While this and prior studies support a mechanism of 'pocket brigade' in which cholesterol would transfer from one pocket to another (NPC2 to the NTD of NPC1 to the SSD of NPC1), the subsequent step involving the transfer of cholesterol to the cytoplasmic leaflet of the endosomal membrane remains completely elusive.

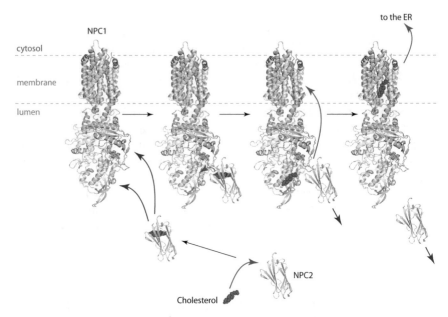

Fig. 6 Putative steps in the mechanism of NPC1/NPC2 mediated cholesterol egress from lyso-somes. A combined picture of models proposed to describe the binding of cholesterol to NPC2 followed by its transfer to NPC1, and subsequently its egress from the lysosome via a yet to be determined mechanism. Conformational changes that occur during the process are not depicted. Also, based on current information, it is unclear whether NPC2 remains bound to NPC1 when cholesterol is transferred to the NPC1 sterol sensing domain, and subsequently, when cholesterol is exported from the lysosome. The model depicts one of the possible scenarios that need to be tested. Black arrows show the sequence of events, red arrows refer to the movement of cholesterol, and purple arrows, to the movement of NPC2. NPC1 and NPC2 are depicted in ribbon representation using the same color scheme as the one used in Fig. 2. Cholesterol is depicted in ball representation in red

In 2017, a crystal structure of NPC1 determined by the Blobel laboratory at 3.3 Å resolution added another player to the picture, the C-terminal luminal domain (CTD) of NPC1 (Fig. 2a) [81]. NPC1 possesses three distinct luminal domains. However, in the absence of a high resolution structure that included the CTD, significant effort has been previously put primarily into determining the roles of the NTD and MLD in the cholesterol transfer from NPC2 to NPC1. Yet, while this crystal structure offered increased resolution of the CTD of NPC1, it lacked the NTD and the TM1 helix of NPC1. Thus, to obtain a more complete structure of NPC1, the crystal structure [81] was aligned to the full-length 2016 Cryo-EM structure [73] that included the missing domains (Fig. 2a). Using the aligned the structures of NPC1, the interactions of the CTD with the NTD were delineated revealing an interaction surface larger than the interaction surface between the NTD and the MLD. These predominantly involved hydrophobic interactions (G910, G911, M912, and G913 of the CTD and V234, T235, and A236 of the NTD) between two distinct loops. In addition, a secondary interface between the CTD and NTD was found to involve primarily polar residues (Q60, E233 of the NTD interacting with L982, Q988 of the CTD, respectively). Furthermore, disruption of the interface between the NTD and CTD of NPC1 via mutagenesis disrupted the transfer of cholesterol from the late endosome to the ER. It was thus suggested that the CTD-NTD interaction may play a critical role in orienting the NTD of NPC1 to facilitate the transfer of cholesterol from NPC2, as well as in modulating the interaction between the NTD and the SSD that would facilitate cholesterol export from the late endosomes [81].

6   Recently Reported Cholesterol Transporters

In addition to NPC1 and NPC2, a number of proteins have been shown or implicated in cholesterol transport [82]. Recent reviews have covered topics ranging from lipoprotein involvement, ABC lipid transporters [83, 84], steroidogenic acute regulatory domain (StARD) and StAR-related lipid transport (START)-domain proteins among others [85–87]. Below are examples of recent findings of proteins involved in cellular cholesterol transport.

1. Sandhu and colleagues reported earlier this year on the involvement of endoplasmic reticulum (ER) proteins termed 'Aster' proteins in cholesterol transport [88] The Aster-A, B and C proteins are structurally similar to StARD and START-domain containing proteins that facilitate cholesterol transport from the plasma membrane to the ER in a non-vesicular mediated process that requires a 'bridge-like' structure to mediate the cholesterol transfer.

2. The protein Patched1 homolog 1 (PTCH1) receptor serves to bind Hedgehog (Hh) proteins and downstream signal transduction including activation of Smoothened (SMO). The regulation of SMO has for some time been known to require cholesterol [89–95]. Most recently, a cryo-EM study provided evidence that PTCH1 has an NPC1-like structural topology, acts as a cholesterol transporter, and alters inner leaflet cholesterol in cells [96]. The latter is reversed by Hedgehog stimulation, suggesting that PTCH1 regulates Smoothened by controlling cholesterol availability [96]. Further structural evidence of PTCH1 and NPC1 similarities was also reported by Qi et al. [97].

While many proteins have been predicted or shown to bind cholesterol, direct evidence of cholesterol transport remains challenging and many transport mechanisms are based on indirect evidence. Nonetheless, combined structural, in vitro and genetic investigations have provided our current understanding of cholesterol transport in the cell.

7   Concluding Remarks and Outlook

Since the observation in 1996 that mutations in NPC1 and NPC2 result in similar NPC disease phenotypes, and may thereby function in a common pathway affecting cholesterol transport [26], the molecular mechanism underlying this process has been gradually unraveling via biochemical, structural and computational efforts (summarized in Fig. 7). Subsequent to the identification of sterol binding sites in the

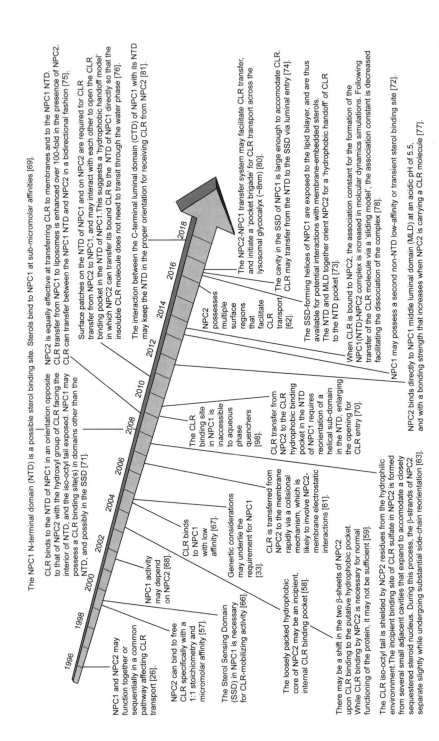

Fig. 7 Developments in the discovery of the molecular mechanism of cholesterol transfer from NPC2 and NPC1. Highlights of major discoveries involving the binding of cholesterol to NPC2 and NPC1, and the proposed mechanisms underlying the transfer of cholesterol from NPC2 and NPC1 during the egress process of cholesterol from lysosomes. In this figure, cholesterol is represented by the three letter abbreviation CLR

two individual proteins, clues have begun to emerge regarding the potential mechanism of cholesterol transfer from NPC2 to NPC1, and the involvement of the different NPC1 domains in this process (summarized in Fig. 6). Notably, the proposed interaction between NPC1 and NPC2 that forms to facilitate cholesterol egress from lysosomes may underlie the common disease phenotype traced to mutations in the two proteins.

However, while it is well accepted that NPC2 and NPC1 play important roles in cholesterol trafficking, several parts of the puzzle are still missing. For example, does cholesterol undergo a conformational change when it is transferred from NPC2 to NPC1? How is NPC2 oriented with respect to the NPC1 luminal domains before and after transferring a cholesterol molecule to NPC1? If cholesterol is transferred from the NTD of NPC1 to the SSD, how does this occur? Is NPC2 released before the cholesterol molecule transfers to the SSD or before it is exported? Most importantly, how cholesterol is exported from NPC1 remains to be shown.

References

1. Yeagle PL. Cholesterol and the cell membrane. Biochim Biophys Acta. 1985;822:267–87.
2. Yeagle PL. Modulation of membrane function by cholesterol. Biochimie. 1991;73:1303–10.
3. Gimpl G, Burger K, Fahrenholz F. Cholesterol as modulator of receptor function. Biochemistry. 1997;36:10959–74.
4. Goluszko P, Nowicki B. Membrane cholesterol: a crucial molecule affecting interactions of microbial pathogens with mammalian cells. Infect Immun. 2005;73:7791–6.
5. Ramprasad OG, Srinivas G, Rao KS, Joshi P, Thiery JP, Dufour S, Pande G. Changes in cholesterol levels in the plasma membrane modulate cell signaling and regulate cell adhesion and migration on fibronectin. Cell Motil Cytoskeleton. 2007;64:199–216.
6. Rosenhouse-Dantsker A, Mehta D, Levitan I. Regulation of Ion channels by membrane lipids. Compr Physiol. 2012;2:31–68.
7. Maxfield FR, van Meer G. Cholesterol, the central lipid of mammalian cells. Curr Opin Cell Biol. 2010;22:422–9.
8. Berg JM, Tymczko JL, Stryer L. Section 26.3. The complex regulation of cholesterol biosynthesis takes place at several levels. In: Berg JM, et al., editors. Biochemistry. 7th ed. New York: W.H. Freeman; 2012. p. 770–9.
9. Afonso SM, Machado RM, Lavrador MS, Quintao ECR, Moore KJ, Lottenberg AM. Molecular pathways underlying cholesterol homeostasis. Nutrients. 2018;10:E760.
10. Zhang J, Liu Q. Cholesterol metabolism and homeostasis in the brain. Protein Cell. 2015;6:254–64.
11. Goedeke L, Fernandez-Hernando C. Regulation of cholesterol homeostasis. Cell Mol Life Sci. 2012;69:915–30.
12. Brown MS, Goldstein JL. A receptor-mediated pathway for cholesterol homeostasis. Science. 1986;232:34–47.
13. Goldstein JL, Dana SE, Faust JR, Beaudet AL, Brown MS. Role of lysosomal acid lipase in the metabolism of plasma low density lipoprotein. Observations in cultured fibroblasts from a patient with cholesteryl ester storage disease. J Biol Chem. 1975;250:8487–95.
14. Rosenbaum AI, Maxfield FR. Niemann-Pick type C disease: molecular mechanisms and potential therapeutic approaches. J Neurochem. 2011;116:789–95.
15. Niemann A. Ein unbekanntes Krankheitsbild. Jahrbuch für Kinderheilkunde, vol. 79. Berlin: Neue Folge; 1914. p. 1–10.

16. Pick L. Der Morbus Gaucher und die ihm ähnlichen Krankheiten (die lipoidzellige Splenohepatomegalie Typus Niemann und die diabetische Lipoidzellenhypoplasie der Milz), vol. 29. Berlin: Ergebnisse der Inneren Medizin und Kinderheilkunde; 1926. p. 519–627.
17. Pentchev PG, Comly ME, Kruth HS, Vanier MT, Wenger DA, Patel S, et al. A defect in cholesterol esterification in Niemann-Pick disease (type C) patients. Proc Natl Acad Sci U S A. 1985;82(23):8247–51.
18. Pentchev PG, Comly ME, Kruth HS, Patel S, Proestel M, Weintroub H. The cholesterol storage disorder of the mutant BALB/c mouse. A primary genetic lesion closely linked to defective esterification of exogenously derived cholesterol and its relationship to human type C Niemann-Pick disease. J Biol Chem. 1986;261(6):2772–7.
19. Pentchev PG, Kruth HS, Comly ME, Butler JD, Vanier MT, Wenger DA, Patel S. Type C Niemann-Pick disease. A parallel loss of regulatory responses in both the uptake and esterification of low density lipoprotein-derived cholesterol in cultured fibroblasts. J Biol Chem. 1986;261(35):16775–80.
20. Vanier MT, Rodriguez-Lafrasse C, Rousson R, Gazzah N, Juge MC, Pentchev PG, Revol A, Louisot P, Type C. Niemann-Pick disease: spectrum of phenotypic variation in disruption of intracellular LDL-derived cholesterol processing. Biochim Biophys Acta. 1991;1096:328–37.
21. Vanier MT, Wenger DA, Comly ME, Rousson R, Brady RO, Pentchev PG. Niemann-Pick disease group C: clinical variability and diagnosis based on defective cholesterol esterification. A collaborative study on 70 patients. Clin Genet. 1998;33:331–48.
22. Fink JK, Filling-Katz MR, Sokol J, Cogan DG, Pikus A, Sonies B, Soong B, Pentchev PG, Comly ME, Brady RO. Clinical spectrum of Niemann-Pick disease type C. Neurology. 1989;39:1040–9.
23. Schiffmann R. Niemann-Pick disease type C. From bench to bedside. JAMA. 1996;276:561–4.
24. Vanier MT, Rodriguez-Lafrasse C, Rousson R, Duthel S, Harzer K, Pentchev PG, Revol A, Louisot P. Type C Niemann-Pick disease: biochemical aspects and phenotypic heterogeneity. Dev Neurosci. 1991;13:307–14.
25. Carstea ED, Polymeropoulos MH, Parker CC, Detera-Wadleigh SD, O'Neill RR, Patterson MC, Goldin E, Xiao H, Straub RE, Vanier MT. Linkage of Niemann-Pick disease type C to human chromosome 18. Proc Natl Acad Sci U S A. 1993;90:2002–4.
26. Vanier MT, Duthel S, Rodriguez-Lafrasse C, Pentchev P, Carstea ED. Genetic heterogeneity in Niemann–Pick C disease: a study using somatic cell hybridization and linkage analysis. Am J Hum Genet. 1996;58:118–25.
27. Carstea ED, Morris JA, Coleman KG, Loftus SK, Zhang D, Cummings C, et al. Niemann-Pick C1 disease gene: homology to mediators of cholesterol homeostasis. Science. 1997;277(5323):228–31.
28. Naureckiene S, Sleat DE, Lackland H, Fensom A, Vanier MT, Wattiaux R, et al. Identification of HE1 as the second gene of Niemann-Pick C disease. Science. 2000;290(5500):2298–301.
29. Davidson CD, Steven UW. Niemann-Pick Type C disease—pathophysiology and future perspectives for treatment. US Neurology. 2010;6:88–94.
30. Crocker AC. The cerebral defect in Tay-Sachs disease and Niemann-Pick disease. J Neurochem. 1961;7:69–80.
31. Jones I, He X, Katouzian F, Darroch PI, Schuchman EH. Characterization of common SMPD1 mutations causing types A and B Niemann-Pick disease and generation of mutation-specific mouse models. Mol Genet Metab. 2008;95(3):152–62.
32. Vanier MT. Complex lipid trafficking in Niemann-Pick disease type C. J Inherit Metab Dis. 2015;38(1):187–99.
33. Pentchev PG. Niemann-Pick C research from mouse to gene. Biochim Biophys Acta. 2004;1685(1-3):3–7.
34. Patterson MC, Walkley SU. Niemann-Pick disease, type C and Roscoe Brady. Mol Genet Metab. 2017;120(1-2):34–7.
35. Adachi M, Volk BW, Schneck L. Animal model of human disease: Niemann-Pick Disease type C. Am J Pathol. 1976;85(1):229–32.

36. Kruth HS, Comly ME, Butler JD, Vanier MT, Fink JK, Wenger DA, et al. Type C Niemann-Pick disease. Abnormal metabolism of low density lipoprotein in homozygous and heterozygous fibroblasts. J Biol Chem. 1986;261(35):16769–74.
37. Liscum L, Faust JR. Low density lipoprotein (LDL)-mediated suppression of cholesterol synthesis and LDL uptake is defective in Niemann-Pick type C fibroblasts. J Biol Chem. 1987;262(35):17002–8.
38. Elleder M, Jirasek A, Smid F, Ledvinova J, Besley GT. Niemann-Pick disease type C. Study on the nature of the cerebral storage process. Acta Neuropathol. 1985;66(4):325–36.
39. Kidder LH, Colarusso P, Stewart SA, Levin IW, Appel NM, Lester DS, et al. Infrared spectroscopic imaging of the biochemical modifications induced in the cerebellum of the Niemann-Pick type C mouse. J Biomed Opt. 1999;4(1):7–13.
40. German DC, Quintero EM, Liang CL, Ng B, Punia S, Xie C, et al. Selective neurodegeneration, without neurofibrillary tangles, in a mouse model of Niemann-Pick C disease. J Comp Neurol. 2001;433(3):415–25.
41. Tanaka J, Nakamura H, Miyawaki S. Cerebellar involvement in murine sphingomyelinosis: a new model of Niemann-Pick disease. J Neuropathol Exp Neurol. 1988;47(3):291–300.
42. Higashi Y, Murayama S, Pentchev PG, Suzuki K. Cerebellar degeneration in the Niemann-Pick type C mouse. Acta Neuropathol. 1993;85(2):175–84.
43. Sarna JR, Larouche M, Marzban H, Sillitoe RV, Rancourt DE, Hawkes R. Patterned Purkinje cell degeneration in mouse models of Niemann-Pick type C disease. J Comp Neurol. 2003;456(3):279–91.
44. Fu R, Yanjanin NM, Bianconi S, Pavan WJ, Porter FD. Oxidative stress in Niemann-Pick disease, type C. Mol Genet Metab. 2010;101(2-3):214–8.
45. Klein A, Maldonado C, Vargas LM, Gonzalez M, Robledo F, Perez de Arce K, et al. Oxidative stress activates the c-Abl/p73 proapoptotic pathway in Niemann-Pick type C neurons. Neurobiol Dis. 2011;41(1):209–18.
46. Lloyd-Evans E, Morgan AJ, He X, Smith DA, Elliot-Smith E, Sillence DJ, et al. Niemann-Pick disease type C1 is a sphingosine storage disease that causes deregulation of lysosomal calcium. Nat Med. 2008;14(11):1247–55.
47. Saez PJ, Orellana JA, Vega-Riveros N, Figueroa VA, Hernandez DE, Castro JF, et al. Disruption in connexin-based communication is associated with intracellular Ca(2)(+) signal alterations in astrocytes from Niemann-Pick type C mice. PLoS One. 2013;8(8):e71361.
48. Byun K, Kim D, Bayarsaikhan E, Oh J, Kim J, Kwak G, et al. Changes of calcium binding proteins, c-Fos and COX in hippocampal formation and cerebellum of Niemann-Pick, type C mouse. J Chem Neuroanat. 2013;52:1–8.
49. Kennedy BE, LeBlanc VG, Mailman TM, Fice D, Burton I, Karakach TK, et al. Pre-symptomatic activation of antioxidant responses and alterations in glucose and pyruvate metabolism in Niemann-Pick Type C1-deficient murine brain. PLoS One. 2013;8(12):e82685.
50. Elrick MJ, Lieberman AP. Autophagic dysfunction in a lysosomal storage disorder due to impaired proteolysis. Autophagy. 2013;9(2):234–5.
51. Cougnoux A, Cluzeau C, Mitra S, Li R, Williams I, Burkert K, et al. Necroptosis in Niemann-Pick disease, type C1: a potential therapeutic target. Cell Death Dis. 2016;7:e2147.
52. Cologna SM, Cluzeau CV, Yanjanin NM, Blank PS, Dail MK, Siebel S, et al. Human and mouse neuroinflammation markers in Niemann-Pick disease, type C1. J Inherit Metab Dis. 2014;37(1):83–92.
53. Zervas M, Dobrenis K, Walkley SU. Neurons in Niemann-Pick disease type C accumulate gangliosides as well as unesterified cholesterol and undergo dendritic and axonal alterations. J Neuropathol Exp Neurol. 2001;60(1):49–64.
54. German DC, Quintero EM, Liang C, Xie C, Dietschy JM. Degeneration of neurons and glia in the Niemann-Pick C mouse is unrelated to the low-density lipoprotein receptor. Neuroscience. 2001;105(4):999–1005.
55. Walkley SU, Suzuki K. Consequences of NPC1 and NPC2 loss of function in mammalian neurons. Biochim Biophys Acta. 2004;1685(1-3):48–62.

56. Kirchhoff C, Osterhoff C, Young L. Molecular cloning and characterization of HE1, a major secretory protein of the human epididymis. Biol Reprod. 1996;54(4):847–56.
57. Okamura N, Kiuchi S, Tamba M, Kashima T, Hiramoto S, Baba T, Dacheux F, Dacheux JL, Sugita Y, Jin YZ. A porcine homolog of the major secretory protein of human epididymis, HE1, specifically binds cholesterol. Biochim Biophys Acta. 1999;1438:377–87.
58. Friedland N, Liou HL, Lobel P, Stock AM. Structure of a cholesterol-binding protein deficient in Niemann–Pick type C2 disease. Proc Natl Acad Sci U S A. 2003;100:2512–7.
59. Ko DC, Binkley J, Sidow A, Scott MP. The integrity of a cholesterol-binding pocket in Niemann-Pick C2 protein is necessary to control lysosome cholesterol levels. Proc Natl Acad Sci U S A. 2003;100:2518–25.
60. Sleat DE, Wiseman JA, El-Banna M, Price SM, Verot L, Shen MM, Tint GS, Vanier MT, Walkley SU, Lobel P. Genetic evidence for nonredundant functional cooperativity between NPC1 and NPC2 in lipid transport. Proc Natl Acad Sci U S A. 2004;101:5886–91.
61. Cheruku SR, Xu Z, Dutia R, Lobel P, Storch J. Mechanism of cholesterol transfer from the Niemann-Pick type C2 protein to model membranes supports a role in lysosomal cholesterol transport. J Biol Chem. 2006;281:31594–604.
62. McCauliff LA, Xu Z, Li R, Kodukula S, Ko DC, Scott MP, Kahn PC, Storch J. Multiple surface regions on the Niemann-Pick C2 protein facilitate intracellular cholesterol transport. J Biol Chem. 2015;290:27321–31.
63. Xu S, Benoff B, Liou HL, Lobel P, Stock AM. Structural basis of sterol binding by NPC2, a lysosomal protein deficient in Niemann–Pick type C2 disease. J Biol Chem. 2007;282:23525–31.
64. Loftus SK, Morris JA, Carstea ED, Gu JZ, Cummings C, Brown A, et al. Murine model of Niemann-Pick C disease: mutation in a cholesterol homeostasis gene. Science. 1997;277(5323):232–5.
65. Patterson MC, Vanier MT, Suzuki K, Morris JA, Carstea E, Neufeld EB, Blanchette-Mackie JE, Pentchev PG. In: Scriver CR, Beaudet AL, Sly WS, Valle D, editors. The metabolic and molecular bases of inherited disease, vol. III. New York: McGraw-Hill; 2001. p. 3611–33.
66. Watari H, Blanchette-Mackie EJ, Dwyer NK, Watari M, Neufeld EB, Patel S, Pentchev PG, Strauss JF III. Mutations in the leucine zipper motif and sterol-sensing domain inactivate the Niemann-Pick C1 glycoprotein. J Biol Chem. 1999;274:21861–6.
67. Ohgami N, Ko DC, Thomas M, Scott MP, Chang CC, Chang TY. Binding between the Niemann-Pick C1 protein and a photoactivatable cholesterol analog requires a functional sterol-sensing domain. Proc Natl Acad Sci U S A. 2004;101(34):12473–8.
68. Ioannou YA. Multidrug permeases and subcellular cholesterol transport. Nat Rev Mol Cell Biol. 2001;2:657–68.
69. Infante RE, Abi-Mosleh L, Radhakrishnan A, Dale JD, Brown MS, Goldstein JL. Purified NPC1 protein. I. Binding of cholesterol and oxysterols to a 1278-amino acid membrane protein. J Biol Chem. 2008;283:1052–63.
70. Kwon HJ, Abi-Mosleh L, Wang ML, Deisenhofer J, Goldstein JL, Brown MS, Infante RE. Structure of N-terminal domain of NPC1 reveals distinct subdomains for binding and transfer of cholesterol. Cell. 2009;137:1213–24.
71. Infante RE, Radhakrishnan A, Abi-Mosleh L, Kinch LN, Wang ML, Grishin NV, Goldstein JL, Brown MS. Purified NPC1 protein: II. Localization of sterol binding to a 240-amino acid soluble luminal loop. J Biol Chem. 2008;283:1064–75.
72. Ohgane K, Karaki F, Dodo K, Hashimoto Y. Discovery of oxysterol-derived pharmacological chaperones for NPC1: implication for the existence of second sterol-binding site. Chem Biol. 2013;20:391–402.
73. Gong X, Qian H, Zhou X, Wu J, Wan T, Cao P, Huang W, Zhao X, Wang X, Wang P, Shi Y, Gao GF, Zhou Q, Yan N. Structural insights into the Niemann-Pick C1 (NPC1)-mediated cholesterol transfer and Ebola infection. Cell. 2016;165(6):1467–78.
74. Li X, Wang J, Coutavas E, Shi H, Hao Q, Blobel G. Structure of human Niemann–Pick C1 protein. Proc Natl Acad Sci U S A. 2016;113:8212–7.

75. Infante RE, Wang ML, Radhakrishnan A, Kwon HJ, Brown MS, Goldstein JL. NPC2 facilitates bidirectional transfer of cholesterol between NPC1 and lipid bilayers, a step in cholesterol egress from lysosomes. Proc Natl Acad Sci U S A. 2008;105:15287–92.
76. Wang ML, Motamed M, Infante RE, Abi-Mosleh L, Kwon HJ, Brown MS, Goldstein JL. Identification of surface residues on Niemann-Pick C2 essential for hydrophobic handoff of cholesterol to NPC1 in lysosomes. Cell Metab. 2010;12:166–73.
77. Deffieu MS, Pfeffer SR. Niemann-Pick type C 1 function requires lumenal domain residues that mediate cholesterol-dependent NPC2 binding. Proc Natl Acad Sci U S A. 2011;108:18932–6.
78. Estiu G, Khatri N, Wiest O. Computational studies of the cholesterol transport between NPC2 and the N-terminal domain of NPC1 (NPC1(NTD)). Biochemistry. 2013;52(39):6879–91.
79. Elghobashi-Meinhardt N. Niemann-Pick type C disease: a QM/MM study of conformational changes in cholesterol in the NPC1(NTD) and NPC2 binding pockets. Biochemistry. 2014;53:6603–14.
80. Li X, Saha P, Li J, Blobel G, Pfeffer SR. Clues to the mechanism of cholesterol transfer from the structure of NPC1 middle lumenal domain bound to NPC2. Proc Natl Acad Sci U S A. 2016;113:10079–84.
81. Li X, Lu F, Trinh MN, Schmiege P, Seemann J, Wang J, Blobel G. 3.3 Å structure of Niemann–Pick C1 protein reveals insights into the function of the C-terminal luminal domain in cholesterol transport. Proc Natl Acad Sci U S A. 2017;114(34):9116–21.
82. Wüstner D, Solanko K. How cholesterol interacts with proteins and lipids during its intracellular transport. Biochim Biophys Acta. 1848;2015:1908–26.
83. Xavier BM, Jennings WJ, Zein AA, Wang J, Lee JY. Structural snapshot of the cholesterol-transport ATP-binding cassette proteins. Biochem Cell Biol. 2018:1–10.
84. Litvinov DY, Savushkin EV, Dergunov AD. Intracellular and plasma membrane events in cholesterol transport and homeostasis. J Lipids. 2018;2018:3965054.
85. Maxfield FR, Iaea DB, Pipalia NH. Role of STARD4 and NPC1 in intracellular sterol transport. Biochem Cell Biol. 2016;94(6):499–506.
86. Soffientini U, Graham A. Intracellular cholesterol transport proteins: roles in health and disease. Clin Sci (Lond). 2016;130(21):1843–59.
87. Du X, Brown AJ, Yang H. Novel mechanisms of intracellular cholesterol transport: oxysterol-binding proteins and membrane contact sites. Curr Opin Cell Biol. 2015;35:37–42.
88. Sandhu J, Li S, Fairall L, Pfisterer SG, Gurnett JE, Xiao X, et al. Aster proteins facilitate nonvesicular plasma membrane to ER cholesterol transport in mammalian cells. Cell. 2018;175(2):514–29.. e20
89. Byrne EFX, Sircar R, Miller PS, Hedger G, Luchetti G, Nachtergaele S, Tully MD, Mydock-McGrane L, Covey DF, Rambo RP, et al. Structural basis of Smoothened regulation by its extracellular domains. Nature. 2016;535:517–22.
90. Cooper MK, Wassif CA, Krakowiak PA, Taipale J, Gong R, Kelley RI, Porter FD, Beachy PA. A defective response to Hedgehog signaling in disorders of cholesterol biosynthesis. Nat Genet. 2003;33:508–13.
91. Huang P, Nedelcu D, Watanabe M, Jao C, Kim Y, Liu J, Salic A. Cellular cholesterol directly activates smoothened in hedgehog signaling. Cell. 2016;166:1176–1187.e14.
92. Huang P, Zheng S, Wierbowski BM, Kim Y, Nedelcu D, Aravena L, Liu J, Kruse AC, Salic A. Structural basis of smoothened activation in Hedgehog signaling. Cell. 2018;174:312–324. e6.
93. Luchetti G, Sircar R, Kong JH, Nachtergaele S, Sagner A, Byrne EF, Covey DF, Siebold C, Rohatgi R. Cholesterol activates the G-protein coupled receptor smoothened to promote Hedgehog signaling. Elife. 2016;5:e20304.
94. Myers BR, Neahring L, Zhang Y, Roberts KJ, Beachy PA. Rapid, direct activity assays for Smoothened reveal Hedgehog pathway regulation by membrane cholesterol and extracellular sodium. Proc Natl Acad Sci U S A. 2017;114:E11141–50.
95. Xiao X, Tang JJ, Peng C, Wang Y, Fu L, Qiu ZP, Xiong Y, Yang LF, Cui HW, He XL, et al. Cholesterol modification of smoothened is required for Hedgehog signaling. Mol Cell. 2017;66:154–162.e10.

96. Zhang Y, Bulkley DP, Xin Y, Roberts KJ, Asarnow DE, Sharma A, et al. Structural basis for cholesterol transport-like activity of the Hedgehog receptor patched. Cell. 2018;175(5):1352–64.e14.
97. Qi X, Schmiege P, Coutavas E, Wang J, Li X. Structures of human Patched and its complex with native palmitoylated sonic hedgehog. Nature. 2018;560(7716):128–32.
98. Liu R, Lu P, Chu JW, Sharom FJ. Characterization of fluorescent sterol binding to purified human NPC1. J Biol Chem. 2009;284:1840–52.

Index

© Springer Nature Switzerland AG 2019
A. Rosenhouse-Dantsker, A. N. Bukiya (eds.), *Direct Mechanisms in Cholesterol Modulation of Protein Function*, Advances in Experimental Medicine and Biology 1135, https://doi.org/10.1007/978-3-030-14265-0